An Identification Guide

Bumble Bees

of North America

An Identification Guide

Bumble Bees

of North America

Paul H. Williams, Robbin W. Thorp,
Leif L. Richardson & Sheila R. Colla

PRINCETON UNIVERSITY PRESS
PRINCETON AND OXFORD

Published by Princeton University Press, 41 William Street, Princeton, New Jersey 08540
In the United Kingdom: Princeton University Press, 6 Oxford Street, Woodstock, Oxfordshire OX20 1TW
press.princeton.edu

ISBN 978-0-691-15222-6
Library of Congress Control Number: 2013945435
British Library Cataloging-in-Publication Data is available

This book has been composed in Minion Pro (main text) and ITC Franklin Gothic (headings and captions)
Printed on acid-free paper ¥
Designed by D & N Publishing, Baydon, Wiltshire, UK
Printed in China
10 9 8 7 6 5 4 3 2 1

CONTENTS

INTRODUCTION

Everybody likes bumble bees. As colorful and familiar visitors to flowers, these insects have long been appreciated by artists, naturalists, and farmers. The eighteenth-century German pioneer of pollination biology, Christian Konrad Sprengel, made observations of their behavior at flowers, and Charles Darwin went on to describe their importance as pollinators at a time when this ecological function had not been widely recognized. In North America, naturalists have been describing their diversity for more than two centuries, but a great deal remains to be done. This guide is aimed at making that easier.

The value of bumble bees as pollinators of wild and cultivated plants is increasingly appreciated. Each year more than a million commercially produced bumble bee colonies are sold around the world. Most of these colonies are used in greenhouses, where their pollination service is worth more than $10 billion annually. The total value of crop pollination by wild bumble bees is far higher than this. Unfortunately, there is convincing evidence that many species of bumble bees in Asia, North America, South America, and Europe are in decline, in part because of accidental introductions of bee diseases by the bumble bee pollination industry. Several of the species described in this guide were commonly encountered through their ranges just 15 years ago, but they are now exceedingly rare. Others have not been seen for years, and one may now be extinct.

North American bumble bees have been the subject of numerous regional guides, but the only comprehensive revision was by Henry J. Franklin in 1913. In recognizing 59 species north of Mexico, Franklin gave full species status to some taxa now considered parts of a species (e.g., "*B. californicus*," a color pattern of *B. fervidus*), but did not describe four taxa now generally agreed to be distinct species with ranges in North America (*B. caliginosus, B. distinguendus, B. franklini,* and *B. vandykei*). H. E. Milliron started a monograph on the bumble bees of the Western Hemisphere in multiple volumes between 1970 and 1973. He recognized only 34 species north of Mexico, but never completed sections on the subgenera *Pyrobombus* or *Psithyrus*. Milliron described a number of the species in this book as subspecies (e.g., he interpreted *B. terricola* as being composed of two subspecies, *B. terricola terricola* and *B. terricola occidentalis*, whereas we recognize these as two separate species). Apparently missing from Milliron's work is *B. franklini*, which he regarded as a variety of *B. occidentalis*, and completely missing is *B. distinguendus*, a species well known in the Old World, but only recently discovered in Alaska. Among the larger regional publications since Franklin are those on the bumble bees of the eastern United States (Plath 1934, Mitchell 1962), western United States (Stephen 1957), California (Thorp, Horning, and Dunning 1983), and eastern Canada (Laverty and Harder 1988). Despite our familiarity with bumble bees and the imperative to address their conservation, in North America there remains a need for a comprehensive modern review of the status of bumble bees and an effective identification manual for those interested in them. Franklin openly expressed uncertainty about his separation of some pairs of closely related North American bumble bee taxa as species, and these issues are still debated in the literature. One reason for our uncertainty about bumble bees is that the color patterns of their hair, which include the most obvious characteristics one might use for identification, can be strikingly variable within species and strongly convergent between them. For example, *B. bifarius* (the so-called Two Form Bumble Bee) occurs throughout western North America in distinctly different red and black color patterns, and at any one site it may bear closer resemblance to other bumble bee species present than to *B. bifarius* color patterns from other parts of its range. To exacerbate the problem, bumble bees

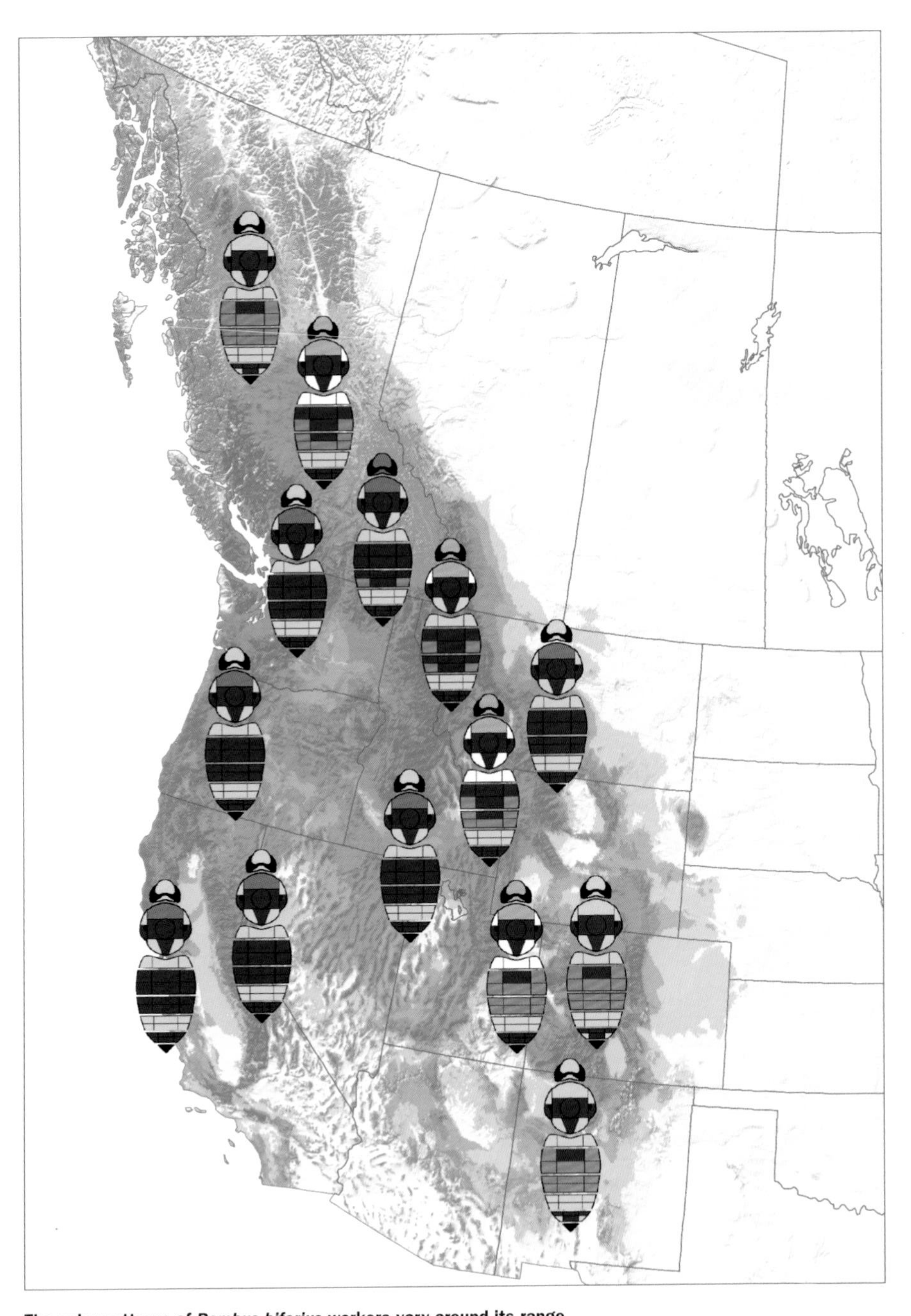

The color patterns of *Bombus bifarius* workers vary around its range.

have relatively few distinctive physical characteristics useful for distinguishing one species from another; the bee systematist Charles Michener has called them "morphologically monotonous." Thankfully, recent molecular analyses have illuminated this debate. This guide presents some new views of species based on molecular phylogenies (family trees) of bumble bees that we hope will improve understanding of these insects.

In addition to describing the distribution and diversity of North American bumble bees, this book is primarily about how to identify them to species. As in some previous guides, we describe hair color patterns in a diagram for each caste (queen or worker), dividing the body into differently colored parts to produce a simplified representation. But for many bumble bee specimens, identification is not possible by use of these color patterns alone. We therefore emphasize the importance of other morphological characteristics, for example the shape of the face and of the male genitalia. Identification of this group of insects is more difficult than is commonly assumed, and experience has shown that there is no substitute—even for experts—for these more subtle characters. That said, we have attempted to make the process of identifying bumble bees as straightforward as we can, stripping away technical language where possible and presenting a system that specialists and beginners alike should be able to use.

BUMBLE BEE DISTRIBUTION AND DIVERSITY

Bumble bees belong to the insect order Hymenoptera, which includes the bees, ants, wasps, and sawflies. Like other similar insect groups such as Diptera (flies), Coleoptera (beetles), and Lepidoptera (butterflies and moths), the Hymenoptera go through a complete metamorphosis between the larval and adult stages of development. Within this group, bees are essentially hairy wasps, from which they diverged more than 100 million years ago. Bees differ from most living wasp lineages in that plant pollen, rather than animal tissue, provides the protein necessary for larval development, and it is thought that bee diversification tracked the adaptive radiation of flowering plants in the late Cretaceous period. Bees differ from wasps in a number of ways, not only in that they have branched body hairs and other adaptations for harvesting and carrying pollen, but also in having an array of tongue morphologies that facilitate feeding from nectar at flowers of different depths.

Plant pollen is the primary protein source for virtually all bees, including bumble bees. LR

There are nearly 20,000 species of bees worldwide, of which just 250 belong to the genus *Bombus*, or bumble bees. While bees in general are most diverse in areas with a warm dry "Mediterranean" climate (e.g., coastal California, South Africa, Chile, and the land around the Mediterranean Sea), bumble bee diversity is greatest in cool temperate and montane situations. They are found throughout the Northern Hemisphere, from Arctic tundra to deserts and subtropical forests, with the greatest diversity in the mountains around the Tibetan plateau. Bumble bees are native in the Southern Hemisphere only in South America, although a few species have been introduced, along with clover, in New Zealand and Tasmania. Forty-six species of bumble bees are found across North America

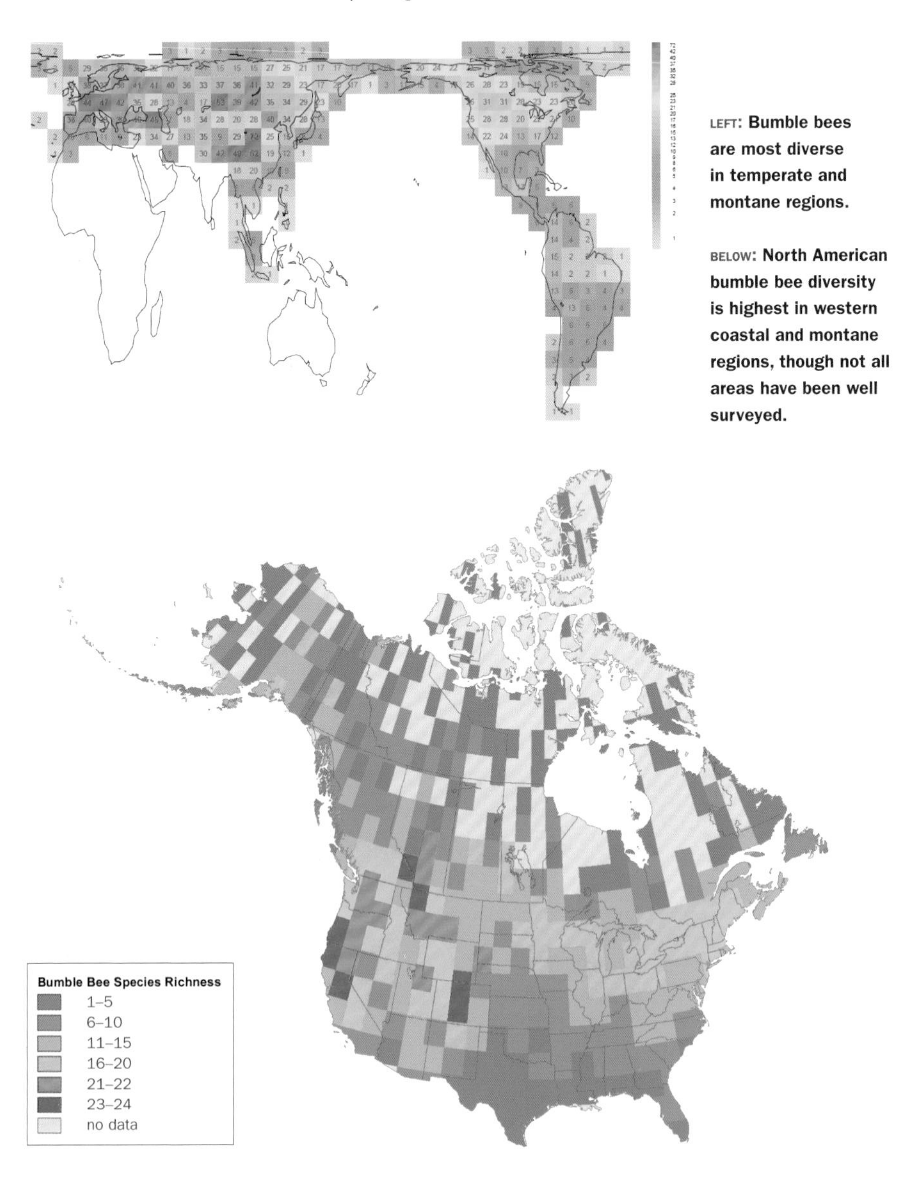

LEFT: Bumble bees are most diverse in temperate and montane regions.

BELOW: North American bumble bee diversity is highest in western coastal and montane regions, though not all areas have been well surveyed.

north of Mexico; although not all areas of the continent have been surveyed equally thoroughly, undoubtedly the greatest diversity occurs in and around the western mountain ranges.

Within their area of occurrence, bumble bees may be found almost anywhere there are suitable flowers. They occupy a wide variety of habitats, reaching peaks of local abundance and diversity

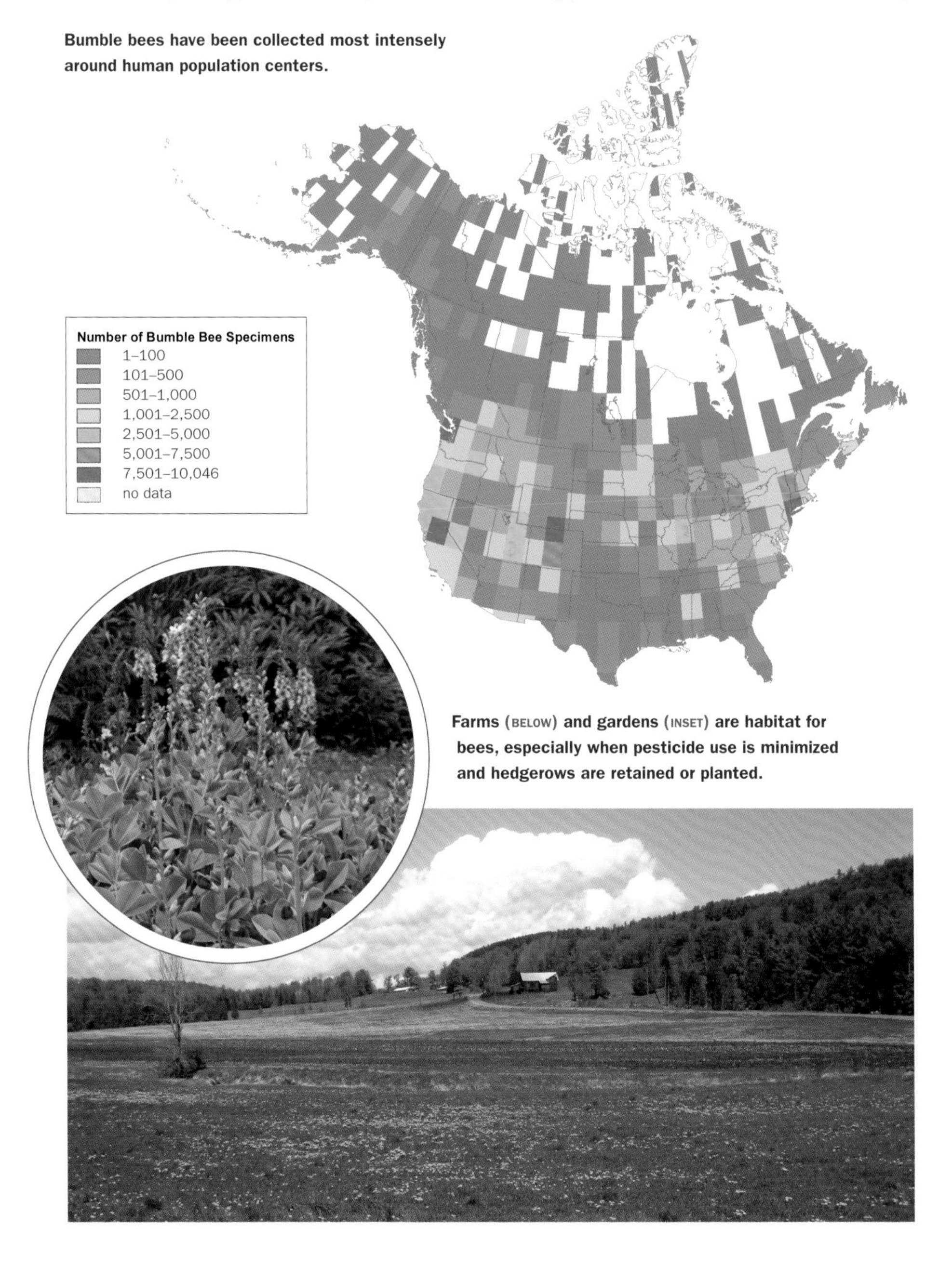

Bumble bees have been collected most intensely around human population centers.

Farms (BELOW) and gardens (INSET) are habitat for bees, especially when pesticide use is minimized and hedgerows are retained or planted.

where a continuous supply of pollen and nectar is available throughout the growing season. Bumble bees occur in mountain meadows, prairies, desert uplands, savannas, agricultural landscapes, gardens, and wetlands, and some species, such as *B. impatiens*, are common even in urban habitats. Forests, especially coniferous forests, usually support relatively fewer bumble bees, but may be attractive to them when spring flowers are in bloom.

— COLONY CYCLE

Bumble bees are social insects. This means that related individuals cooperate to forage for food, rear offspring, and defend their nests. Queens, workers, and males perform different functions within the colony, and there is also specialization within the worker caste. There is evidence of communication between and within groups. As social insects, bumble bees are similar to their close relatives the honey bees, but with an important exception: honey bee colonies can persist with the same queen for years, depending on stored resources during the winter months spent inside the nest, whereas bumble bee colonies die at the end of each growing season, with new ones founded each year. (There is at least one exception to this rule from the American tropics, where climate does not curtail the bees' access to resources as it does in the temperate latitudes.)

The bumble bee life cycle can be thought of as starting in spring (A), when mated, overwintered queens emerge from hibernation to begin the business of founding a colony. At this time, the large, brightly colored bees are a familiar sight as they gather nectar and pollen from the few flowering

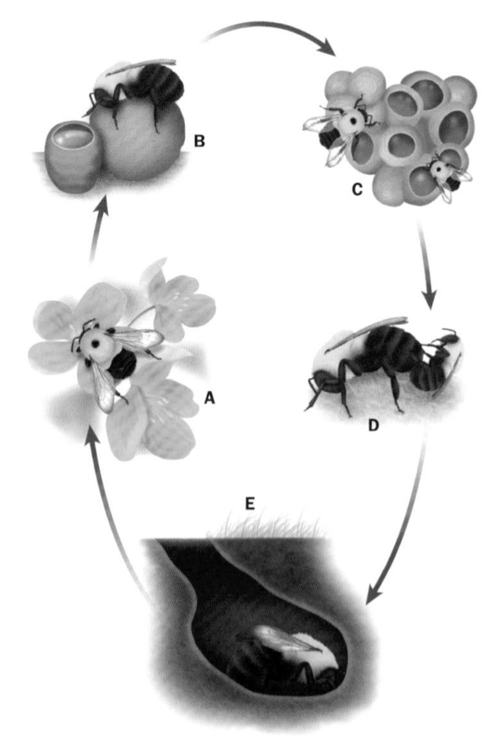

ABOVE: Bumble bees are found in a variety of habitats, including urban centers. LR
RIGHT: Bumble bee life cycle (A–E). See text for details of each stage. AS

A *B. griseocollis* queen in her nest (exposed) before the first batch of workers emerge. LR

plants in an otherwise quiet landscape, including willows, rhododendrons, and ephemeral spring wildflowers. Bumble bees are not warm blooded, but they can maintain a relatively consistent body temperature regardless of the ambient temperature, generating heat by shivering their thoracic flight muscles, so they can be active in the cool, wet weather of early spring. The queen searches for a suitable nest site, flying low over the ground and repeatedly landing to investigate holes in the ground. Bumble bees do not dig their own nest cavities, relying instead on abandoned rodent dens, open grass tussocks, hollow logs, and aboveground manmade structures. Many species choose different nesting substrates, and some have more specific habitat requirements (e.g., those that nest only aboveground in open herbaceous habitats).

When the queen has located a nest site (B), she constructs a wax honeypot for nectar storage. She lays her first clutch of eggs on a mass of pollen moistened with nectar in a small wax cup known as a brood clump. After hatching, the young larvae feed on this pollen, and the queen alternates between incubating the larvae and foraging for more food. This is a vulnerable time in the life of a colony, as vagaries of weather and food—not to mention aggressive interactions with other bumble bee queens in search of nest sites—may determine the fate of the queen and her eggs. Bumble bee species employ two distinct behaviors to provide larvae with pollen. The "pocket-making" species place lumps of pollen into wax pockets attached to the base of the brood clump, and larvae feed together from this supply. The "pollen storers" feed their larvae directly, and larvae leave the brood clump as they grow. Each of these behaviors is associated with one of the two principal evolutionary lineages of bumble bees. Larvae feed for about two weeks before spinning a silk cocoon and pupating for another two weeks. The adult bees that emerge from these pupae are females who will not, in most cases, produce their own offspring, living instead as workers in their mother's colony (C). The queen now stays at home laying eggs, and workers forage for resources, tend new clutches of eggs and larvae, regulate the nest temperature, and defend the nest.

It is in the early stages of development that the colony may be attacked by cuckoo bumble bees. These members of the subgenus *Psithyrus* were long thought to be a group separate from the "true" bumble bees, but we now understand them to be part of the same evolutionary lineage as the rest of the genus. Cuckoo bumble bees do not forage for pollen or found their own colonies; instead, they enter the nests of other species, sometimes kill the queen, and overcome the workers through a combination of aggression and pheromones. The workers then rear the

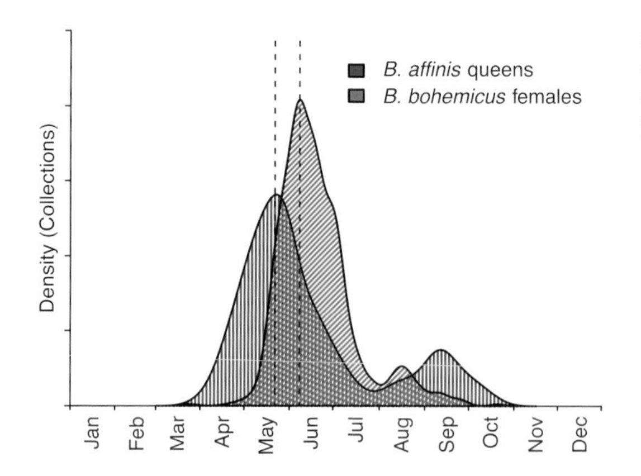

Cuckoo bumble bees, such as *B. bohemicus*, emerge later than their hosts and attack recently founded colonies.

offspring of this usurper, which will be exclusively males and females (no workers). Timing of the attack is critical—to succeed, a social parasite must invade a host nest large enough to raise as many of its offspring as possible, but not so large that the workers will kill it. Cuckoo bumble bees possess a range of adaptations related to their life history, but they lack some of the morphology associated with other bumble bees, such as the pollen-carrying corbiculae of the hind tibiae in females. Cuckoo species often attack a broad range of host species, but some specialize in attacking the members of just one species or subgenus (e.g., *B. variabilis* has been documented as a parasite only in the nests of *B. pensylvanicus*). This social parasitism can be quite common, but cuckoo bumble bee populations must be smaller than those of their hosts, and population trends are constrained by those of their hosts. Many species are now of conservation concern.

Bumble bee colonies grow quickly as successive broods of workers are produced and more floral resources become available. At some point in summer, the colony switches over to the production of males and new queens. This switch is thought to be related to the age of the queen and the size of the colony, though it is not well understood. As with other Hymenoptera, sex determination in bumble bees is controlled by a system known as *haplodiploidy*, in which fertilized eggs (which are diploid, with different genetic material from both parents) develop into female adults and unfertilized eggs (haploid, containing only DNA from their mothers) into males. The queen thus produces males simply by laying eggs not fertilized by the sperm she has stored in her body since mating the previous fall. Unfortunately, in populations with low genetic diversity, fertilized eggs may also develop into males that develop and behave as normal, but are sterile. This reduces a colony's output of reproductive females while introducing males to the population that may mate with but cannot fertilize queens. How fertilized eggs become queens instead of workers is not fully understood, but it is probably associated with larval diet and possibly exposure to queen pheromones.

Adult males do not forage for the colony (with rare exceptions), leaving the nest after emergence to feed at flowers and search for mates (D). In some species, males patrol for queens and advertise their presence with pheromones. In other species, the males occupy pheromone-scented perches while waiting for queens to fly past. The latter (e.g., *B. griseocollis*) tend to have greatly enlarged eyes, a trait that can be useful in identification. Newly emerged queens leave the colony to feed during the day, often returning at night. They eat a great deal of pollen and nectar,

LEFT: **Adults tending the brood cells of a *B. impatiens* colony.** LR

BELOW: ***B. huntii* mating pair.** NS

building fat reserves that will carry them through a winter of hibernation (E). Bumble bee queens usually mate once with only one male. They then search for a suitable overwintering site and enter a period of torpor. Relatively little is known about where bumble bee queens spend the winter, but they have been reported to use burrows of other animals and to excavate holes in loose dirt or in debris such as that of compost piles. The new queen having reproduced, the colony declines, with males, workers, and the old queen dying before winter.

INTERACTIONS WITH PLANTS

Bumble bees have specific habitat requirements for nesting and overwintering, and a third aspect of habitat, forage, is even more important. Relative to many other bees, bumble bees have a long period of summer activity, and with only modest resources stored in the nest, pollen and nectar must be available continuously. These animals collect resources from a patchy environment and return to one central point—the nest—to consume them. The distance bees are capable of traveling on a foraging trip is a critical aspect of their foraging ecology, as are the energetics of flight for a resource-laden bee. Bumble bee species vary in their ability to travel the landscape, and the farthest they have been demonstrated to fly from the nest is about 10 km. Whether or not they often fly this far, to be successful foragers they must return with more resources than they left with, the cost of the flight included. For this reason, most resource acquisition is likely to take place considerably closer to the nest.

Because bees eat pollen, their evolutionary history is entwined with that of flowering plants, and they are well known as pollinators. Bumble bees possess a number of traits that make them effective pollinators of both wild and cultivated plants. They are classic generalist foragers, capable of working a wide variety of plants for their resources. Although neither they nor any other bees deliberately pollinate flowers, pollen sticks easily to their copious hair and is transferred effectively between flowers. An individual worker often visits just one species of plant at a time, minimizing pollen transfer to other, unreceptive plants. Bumble bees can learn complex foraging tasks such as working the flowers of closed gentians and pea family plants. Unlike most other bees (including honey bees), they often "buzz" flowers by vibrating their flight muscles. This shaking helps them extract pollen from the anthers and greatly facilitates pollination for certain plants. In fact, many plants are adapted to dispense pollen when vibrated, and some even require this. Plants that benefit from "buzz" pollination include those in the nightshade (Solanaceae), rose (Rosaceae), heath (Ericaceae), and melastome (Melastomataceae) families. Numerous crop plants (e.g., blueberry, cranberry, tomato, and kiwi) have improved yields when buzzed by pollen-collecting bumble bees, which is why commercial bumble bee colonies, rather than honey bee hives, are maintained in many greenhouses where crops are grown.

It is common for multiple species of bumble bees to be present in an area, placing them in potential competition with each other for resources. Foraging niches are usually partitioned to some extent by variation in tongue length, and bee species can often be divided roughly into artificial groups by tongue length. Species with long tongues (including, for example, *B. fervidus* and *B. pensylvanicus*) can reach nectar in tubular flowers with long corollas, while those with short tongues (e.g., *B. morrisoni* and *B. terricola*) cannot, and are most efficient at harvesting nectar from smaller flowers. Some short-tongued species circumvent this limitation by biting holes in the base of long flower corollas, allowing them direct access to nectaries. This "nectar robbing" behavior is especially common in species of the subgenus *Bombus s. str.*, in which females' mouthparts seem to have adaptations for cutting. Nectar robbing opens a shortcut to harvesting nectar resources, and many other bees will subsequently switch from typical "legitimate" foraging behaviors to secondary nectar robbing. This can have negative consequences for plant reproduction, and plants have a range of floral traits that prevent bees from nectar robbing. Interestingly, however, research has demonstrated that in many cases the reproductive consequence of nectar robbing for plants is neutral or even positive.

A number of excellent publications detail the life history and biology of bumble bees. A list of resources is provided on page 203.

Nectar robbing holes left in flowers of Squirrel Corn (*Dicendra canadensis*) by queen bumble bees. LR

OBSERVING BUMBLE BEES

Properly identifying a fast-moving bumble bee to species level requires a bit of practice, patience, and some tried-and-true techniques. While a foraging bumble bee is quite docile when busy gathering food from flower to flower, female bumble bees can and will sting if defending their nest or trapped (e.g., under foot, in clothing). Unlike honey bees, which die if they use their strongly barbed sting, bumble bees can often withdraw their weakly barbed sting from skin and use it repeatedly until they are able to escape. However, by being cautious, you will be able to observe bumble bees closely in nature without getting stung.

Bumble bees can be found throughout the year, depending on the site and the activity patterns of local species. Individuals are most often spotted while foraging on flowers or nest-searching near the ground. When flowers are inconspicuous and overgrown (e.g., *Rubus*), bumble bees are sometimes heard before they are seen. Habitats providing the best chance of seeing a variety of species include the following:

- Farms and gardens with a diversity of flowering crops and herbs
- Hayfields, roadsides, ditches, and windbreaks with good abundance and diversity of "weedy" flowering plants such as *Trifolium, Melilotus, Medicago*, and *Vicia*
- Wetlands, especially open habitats like fens and bogs
- Wet meadows, open grasslands, and old fields
- Hardwood forests, when spring ephemerals are in bloom and when queens are nest searching
- Montane meadows
- Urban parks and gardens

Open areas surrounded by woodland are particularly good for collecting bees. SCO

Foraging bumble bees can be placed in a vial while they are distracted. SCO

While bumble bees have been known to forage at temperatures near freezing (32 °F, 0 °C), it is best to search for bumble bees at temperatures ranging from 59 to 86 °F (15–30 °C). If the climate is suitable, they will forage from dawn to dusk. In warm climates, bumble bees are more active in the mornings and evenings. Precipitation-free days with low wind speeds are ideal. Although bumble bees do occasionally forage under adverse conditions, short periods of inclement weather make for more difficult collecting.

If you are interested in learning how to identify bumble bees, you can do so in numerous ways. In some cases, it is possible to identify bumble bees without killing them, but specimens you can examine closely are usually required. Many species have similar color patterns, making correct identification difficult, and you will have to use other morphological characters (see identification key). However, with practice, especially in regions with few species, many bumble bees can be identified relatively easily, and good close-up images can provide enough detail for you to identify them properly. Make sure you get shots of the bee's side (lateral) and face and from the back (dorsal). This is especially easy on cool mornings in the spring and autumn, when the bees are slow moving and trying to warm up inside or underneath flowers. To slow bumble bees down for photos, capture, chill on ice in a portable cooler, pose on the flowers where they were seen or on a substrate of choice, take the photo, and release.

There are many ways to catch bumblebees, and most entomologists have their own favorite techniques and tools. In general, catching bees while they forage on flowers with an insect net is best. Insect nets can be purchased from numerous distributors (e.g., BioQuip, Rose Entomology). There are two main ways to catch bumble bees using an insect net. You can hold your insect net over the bee on the flower, the tip of the net held toward the sky, and wait for the bee to fly upward. Alternatively, you can swipe at the bee with your net, twisting your wrist midair to trap the bee at the tip of the net (although this method can damage the plant). Once the bee is in the net, you can place it in a vial. Empty clear plastic water bottles, pierced for breathing holes and containing dry absorbent paper for the bees to hang onto, are excellent. If you use a clear vial, you can photograph the bee from numerous angles and release it, or if you intend to kill and keep the specimen in an insect collection, you can place it in a cooler with ice or ideally in a freezer overnight. In some cases you can even trap bees directly on a flower by using a jar, without a net.

A variety of traps (e.g., pan traps and blue vane traps) can also be used, especially if a quantitative survey is being made. However, traps filled with liquids, such as propylene glycol or soapy water, result in soggy, bedraggled-looking specimens with matted hair that may be more difficult to identify. After they have been pinned and washed, preferably with alcohol, these specimens need to be fluffed up with a hair dryer or soft paint brush.

Once you have your specimens, you can preserve them on insect pins and keep them in a pest-proof insect pinning box, stored in a cool, dry location. Before they dry out, bees should be pinned through the right-hand side of the thorax (i.e., through the scutum), about two-thirds the way up the pin. Make sure to properly label the specimen with the collection date, collector's name, and location, ideally with a latitude and longitude taken from a GPS unit. It can help with preservation to keep the box in a sealed plastic bag in a cool dry place.

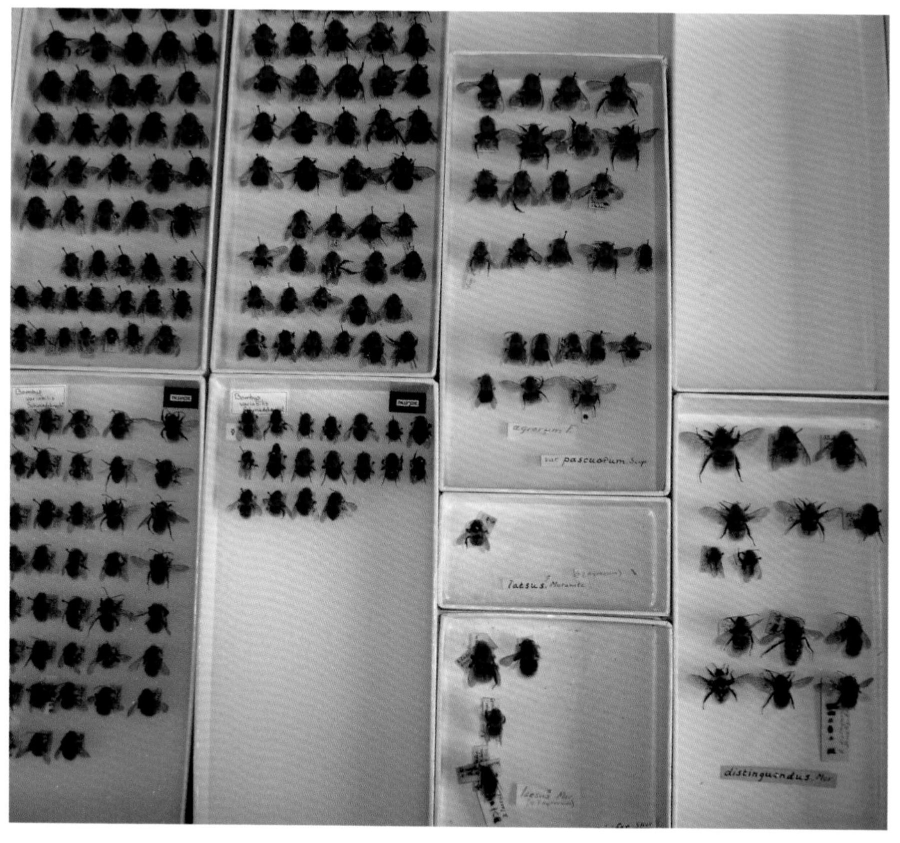

Bumble bees pinned, labeled, and stored. SCO

ATTRACTING BUMBLE BEES

There are several easy ways to attract a variety of bumble bees and other interesting pollinators to your garden. Not only will you benefit by being able to observe and learn to identify these animals from day to day in your own backyard, but you will also be helping them by providing food and shelter, precious resources in our increasingly altered landscapes.

— FOOD SOURCES

Bumble bees feed from pollen- and nectar-rich flowers throughout the spring and summer. Spring queens must visit early-flowering plants to build up energy stores before they can start a new colony. Later in the summer, workers need an abundance of flowers to bring to the growing colony enough resources to produce the next generation of queens and males.

When planting for bumble bees, select plants with different blooming periods and different shapes and colors. The longer-tongued bumble bee species prefer flowers with long corollas such as Beardtongues (*Penstemon* spp.) and Larkspurs (*Delphinium* spp.). Bumble bees with shorter

Worker bumble bees foraging on *Angelica archangelica* in a Yukon garden. SC

LEFT: **Bumble bees prefer plants that provide ample nectar and pollen.** AB

BELOW: **Pollinator gardens like the West Pond Butterfly and Hummingbird Garden in Davis, CA, provide forage in otherwise unsuitable urban environments.** GZ

tongues occasionally nectar rob from flowers with long corollas, but they generally prefer shallow flowers such as Goldenrods (*Solidago* spp.) and Sunflowers (*Helianthus* spp.). By planting a variety of flowering plants (see Forage Plant List) that will bloom throughout the spring, summer, and fall, you will provide several species of bumble bees the nutrients they require to survive from generation to generation.

NEST SITES

Providing nesting resources for bumble bee colonies is an important but often forgotten part of encouraging populations. Some species require tufts of long grass or hay to form protected nests aboveground. For these species, it is beneficial to leave some grassy areas unmowed for a season, or to leave piles of hay.

Other bumble bee species nest in preexisting burrows and crevices. These nests can be in old rodent burrows, in abandoned bird nests, between cinder blocks, under fire hydrants, or even inside abandoned furniture. The best way to encourage these species is to allow them to nest where they find a suitable site without disrupting them. You can also place wooden bumble bee nest boxes with some cotton (upholsterer's type, not medical grade cotton) outside to try to encourage queens to establish nests, but these boxes generally have low occupancy rates.

OVERWINTERING SITES

Mated queens spend the winter in rotting logs, loose dirt, or mulch until emerging in the spring to found a colony. Frequently, people come across overwintering queens when preparing their garden for spring plantings. If possible, provide an area with these substrates, and if you come across a queen while digging, try to bury her back underground without disturbing her.

BUMBLE BEE FORAGE GUIDE BY ECOREGION

(* INDICATES PREDOMINANTLY NON-NATIVE)

In the species accounts, we use quotes around "*Aster*" and "*Epilobium*" because, although these genera have been split taxonomically, bumble bees forage on all the daughter genera. We cannot tell which genus a collector intended when recording the name of a plant from which a bumble bee was collected.

TUNDRA/TAIGA

Yukon Penstemon (*Penstemon gormanii*). SC

- Yukon Penstemon (*Penstemon gormanii*)
- Arctic Lupine (*Lupinus arcticus*)
- Horned Dandelion (*Taraxacum cerotophorum*)
- Purple Saxifrage (*Saxifraga oppositifolia*)
- Northern Labrador Tea (*Rhododendron tomentosum*)
- Lapland Rosebay (*Rhododendron lapponicum*)
- Lowbush Blueberry (*Vaccinium angustifolium*)
- Cloudberry (*Rubus chamaemorus*)
- Sweet Clovers (*Melilotus* spp.*)
- Vetches (*Vicia* spp.*)
- Willowherbs and Fireweeds (*Chamerion* and *Epilobium* spp.)
- Berry shrubs such as blueberry, bilberry (*Vaccinium* spp.)
- Clovers (*Trifolium* spp.*)

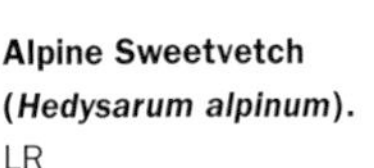

Alpine Sweetvetch (*Hedysarum alpinum*). LR

Wild Bergamot (*Monarda fistulosa*). LR

Goldenrod (*Solidago* sp.). LR

BOREAL FORESTS

- Mat Vetch (*Vicia americana*)
- Yellow Avens (*Geum aleppicum*)
- Fleabane (*Erigeron* spp.)
- Cylindrical Blazing Star (*Liatris cylindracea*)
- Early Buttercup (*Ranunculus fascicularis*)
- Narrow-leaved Vervain (*Verbena simplex*)
- Prairie Smoke (*Geum triflorum*)
- Wild Bergamot (*Monarda fistulosa*)
- White False Indigo (*Baptisia alba*)
- Wild Columbine (*Aquilegia canadensis*)
- Goldenrods (*Solidago* spp.)
- Prairie Clovers (*Dalea* spp.)
- Sweet Clovers (*Melilotus* spp.*)
- Hollyhocks (*Alcea* spp.*)
- Vetches (*Vicia* spp.*)
- Hyssops (*Agastache* spp.)
- Plume Thistles (*Cirsium* spp.*)
- Clovers (*Trifolium* spp.*)
- Apples (*Malus* spp.)

Apple (*Malus* sp.). LR

ABOVE: Purple-flowering Raspberry (*Rubus odoratus*). LR
RIGHT: Larkspur (*Delphinium* sp.). LR

NORTHWEST FORESTED MOUNTAINS

- Wild Rose (*Rosa nutkana*)
- Thimbleberry (*Rubus parviflorus*)
- Western Monkshood (*Aconitum columbianum*)
- Milkvetch (*Astragalus* spp.)
- Larkspur (*Delphinium* spp.)
- Lupines (*Lupinus* spp.)
- Beardtongues (*Penstemon* spp.)
- Fireweed (*Chamerion angustifolium*)
- Plume Thistles (*Cirsium* spp.)
- Rabbitbrush (*Chrysothamnus* spp.)
- Sweet Clovers (*Melilotus* spp.*)
- Ragworts and Groundsels (*Senecio* spp.)
- Clovers (*Trifolium* spp.*)
- Phacelias (*Phacelia* spp.)
- Manzanita (*Arctostaphylos* spp.)
- Salix (*Salix* spp.)

ABOVE RIGHT: Fireweed (*Chamerion angustifolium*). LR
RIGHT: Plume Thistle (*Cirsium muticum*). LR

MARINE WEST COAST FORESTS

- Serviceberry (*Amelanchier alnifolia*)
- Wood Rose (*Rosa gymnocarpa*)
- Thimbleberry (*Rubus parviflorus*)
- Snowberry (*Symphoricarpos alba*)
- Puget Balsamroot (*Balsamorhiza deltoidea*)
- Lupines (*Lupinus* spp.)
- Spiderplants (*Cleome* spp.)
- Sunflowers (*Helianthus* spp.)
- Goldeneyes (*Viguiera* spp.)

Lupine (*Lupinus* sp.). KKG

EASTERN TEMPERATE FORESTS

- Serviceberry (*Amelanchier* spp.)
- Chokecherry (*Prunus virginiana*)
- Lowbush Blueberry (*Vaccinium angustifolium*)
- Harebell (*Campanula rotundifolia*)
- Turtlehead (*Chelone glabra*)
- Joe-pye Weed (*Eupatorium maculatum*)
- Wild Bergamot (*Monarda fistulosa*)
- Obedient Plant (*Physostegia virginiana*)
- Honeysuckle (*Lonicera* spp.)
- Clovers (*Trifolium* spp.*)
- Blazing Stars (*Liatris* spp.)
- Prairie Clovers (*Dalea* spp.)
- Goldenrods (*Solidago* spp.)
- Rosinweeds (*Siphium* spp.)
- Milkweeds (*Asclepias* spp.)
- Plume Thistles (*Cirsium* spp.*)
- Cassias (*Cassia* spp.)
- Salix (*Salix* spp.)
- Impatiens (*Impatiens* spp.)
- Crocus (*Crocus* spp.)
- Dicentra (*Dicentra* spp.)
- Kalmia (*Kalmia* spp.)
- Solanum (*Solanum* spp.)

ABOVE RIGHT: **Serviceberry (*Amelanchier* sp.).** LR
RIGHT: ***Prunus* sp.** LR

TOP LEFT: **Turtlehead (*Chelone glabra*).** LR
TOP RIGHT: **Joe-pye Weed (*Eupatorium maculatum*).** LR
ABOVE LEFT: **Willow (*Salix* sp.).** LR
ABOVE RIGHT: ***Impatiens capensis*.** LR
LEFT: ***Crocus* sp.** LR

Kalmia polifolia. LR

Solanum carolinense. LR

GREAT PLAINS

- New Jersey Tea (*Ceanothus americanus*)
- Milkweed (*Asclepias* spp.)
- Wild Indigo (*Baptisia australis*)
- Larkspur (*Delphinium* spp.)
- Prairie Clovers (*Dalea* spp.)
- Narrow-leaved Purple Coneflower (*Echinacea angustifolia*)
- Blazing Star (*Liatris* spp.)
- Azure Blue Sage (*Salvia azurea*)
- Bee Balms (*Monarda* spp.)
- Sweet Clovers (*Melilotus* spp.*)
- Clovers (*Trifolium* spp.*)
- Goldenrods (*Solidago* spp.)
- Vetches (*Vicia* spp.*)
- "Asters" (*Symphyotrichum* spp.)
- Thistles (*Carduus* spp.*)
- Hollyhocks (*Alcea* spp.)
- Hyssops (*Agastache* spp.)
- Vervains (*Verbena* spp.)
- Sunflowers (*Helianthus* spp.)

Blazing Star (*Liatris* sp.). VM

TOP LEFT: **Dutchman's Breeches (*Dicentra cucullaria*).** LR
TOP RIGHT: **Sunflower (*Helianthus* sp.).** LR
LEFT: **Lupine (*Lupinus* sp.)**

NORTH AMERICAN DESERTS AND SOUTHERN SEMIARID HIGHLANDS

- Fairy Duster (*Calliandra eriophylla*)
- Desert Ironwood (*Olneya tesota*)
- Mojave Lupine (*Lupinus sparsiflorus*)
- Parry's Beardtongue (*Penstemon parryi*)

MEDITERRANEAN CALIFORNIA

- California Buckeye (*Aesculus californica*)
- Hairy Ceanothus (*Ceanothus oliganthus*)
- California Buckwheat (*Eriogonum fasciculatum*)
- Black Sage (*Salvia mellifera*)
- California Milkweed (*Asclepias californica*)
- Short-spike Hedge Nettle (*Stachys pycnantha*)
- Lupines (*Lupinus* spp.)
- Sunflowers (*Helianthus* spp.)

MAPS AND SEASONAL ACTIVITY

Where and when are the different bumble bee species active? Bumble bees have been a popular research focus of North American students, naturalists, and scientists for more than a century; as a result, thousands of pinned specimens, each with a label describing the details of its collection and identity, can be found in the scientific collections of museums, universities, and government institutions. A recent focus of these institutions has been the digitizing of collections in order to make the data accessible to scientists and the general public. We assembled a database of more than 250,000 digitized, georeferenced bumble bee collections to interpret species distribution, seasons of activity, and food preferences. The data come from 120 museums, universities, and private collections from North America and beyond, although several prominent collections are not included because digitized specimen data were unavailable. A small number of records come from online photo-based citizen science projects and from scientific publications. Collection dates range from 1844 to the present, with most data coming from recent decades. Numbers of records per species range from just 20 (for *B. distinguendus*, a recently discovered Alaskan occurrence of a predominantly Old World species) to more than 23,000 (for the Common Eastern Bumble Bee, *B. impatiens*). Collections are generally most thorough in the vicinity of major academic institutions and population centers. Our dataset shows areas of the continent that are undersurveyed for bumble bees, including the southeastern United States and northern Canada, especially northern Quebec. (See map of bumble bee specimen richness on page 11).

Maps are displayed at a scale of 1:33,000,000 (using the Lambert Conformal Conic projection). The maps show state or provincial boundaries, major water bodies and mountain ranges, and other features. Bumble bee specimens collected before 1996 are represented by orange dots and those collected between 1996 and 2012 by red dots. In some cases we accepted data we believed were accurate but could not verify. These records are marked with a question mark. A few species have been introduced far outside their known historical range and these are marked with a "+". Climatic suitability for each species is displayed in green beneath the symbols for the actual bee records. Suitability for each species was modeled using the Maxent procedure (www.cs.princeton.edu/~schapire/maxent/), which combines georeferenced bumble bee presence data (i.e., records of a species' occurrence, but no information on where it does not occur) with climatic data, to predict areas where the species is most likely to occur. We used 19 climate variables derived from monthly precipitation and temperature data from weather stations around the world (www.worldclim.org/). The resulting climatic suitability models have a spatial resolution of 1 km^2 and are depicted in three intensities of green, with the darkest representing areas with greatest probability of species' occurrence. It is important to understand that for each species, the distribution model uses all available bee records, regardless of collection date, even though some species' ranges have changed over time, because the distribution changes are believed to result from adding new constraints, such as new pathogens or loss of food plants, not from climate change. For example, our map for *B. affinis* predicts a broad range of suitable climate across the eastern United States and southern Canada, but nearly all recent records are just from the Midwest.

The species accounts include histograms or bar graphs of seasonal activity patterns, for each caste (for parasitic species just the two sexes are shown). This information is drawn from the same database used to make the maps. In most cases, the graph shows gradual emergence of overwintered queens in spring, then the appearance of workers, and finally emergence of males and new queens.

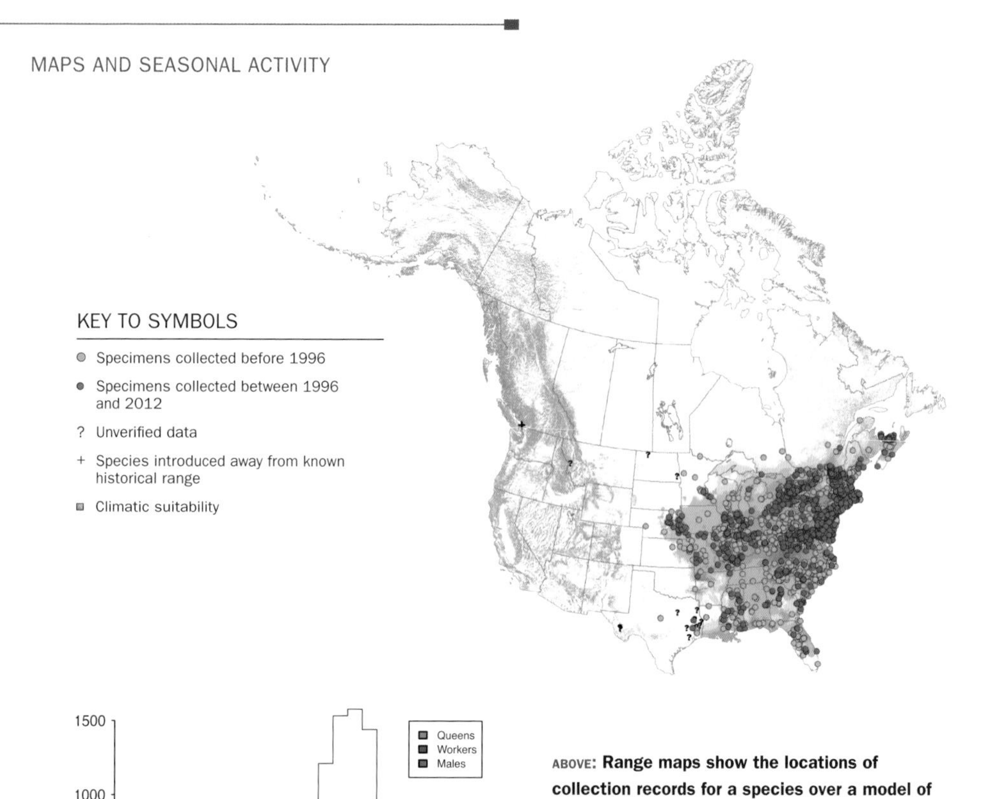

ABOVE: **Range maps show the locations of collection records for a species over a model of the species' climatic range.**

LEFT: **Histograms show seasonal activity patterns of queens (green), workers (blue), and males (red) from collection records.**

Data from collections provide a window into the timing of events for each species, but should be interpreted with some caution, mainly because the information comes from pinned insect specimens, which aren't a perfect representation of nature. For example, for some species, the number of queens and workers seems small relative to the number of males. This is not because of genuine biological differences in production of the different castes, but because some collections report only the sex of the bee, not its caste. In these cases, a specimen labeled "female" could be either a queen or a worker, but it is left out of our seasonal activity graphs altogether. Also, many bumble bee species occur in a range of climates at different latitudes, longitudes, and elevations, and activity periods vary accordingly. The figures include all records for a species, so this variation is obscured. Finally, data from museum collections appear to overemphasize numbers of queens in spring while underestimating their numbers in late summer and fall. This might be because queen bumble bees are one of the only insects on the wing in early spring, but are easily overlooked among the many other bees later in the year, when particular individuals may be flying for only a short period.

BUMBLE BEE DECLINE AND CONSERVATION

Drastic declines of bee populations have been making newspaper headlines in recent years. Reductions in managed honey bee colonies, an introduced species that pollinates many important agricultural crops, have increased awareness about how much we rely on bees in general for their ecosystem services. Our native bumble bees are also important pollinators of a variety of food crops and have recently been found to be in decline in the wild. Bumble bees excel at pollinating many crops grown in greenhouses such as peppers, cucumbers, and tomatoes. They are also efficient pollinators of early blooming field crops such as blueberries and cranberries, and of plants requiring buzz pollination (in which the bee grabs onto the anthers and uses its flight muscles to rapidly vibrate, dislodging the pollen), such as eggplants. Ecologically, bumble bees are also important as pollinators of woodland spring ephemerals and montane meadow plants, as well as many of the native plant species that make up the diverse ecosystems found throughout temperate North America. The loss of bumble bees at a site may cause cascading effects on native plant populations along with birds and mammals that rely on pollinated plants for food and shelter.

Decline of some North American bumble bee species was first identified by Robbin Thorp during his annual surveys in Oregon and California, which began in 1998. The Franklin Bumble Bee (*Bombus franklini*) was once a common bumble bee throughout its small range but has become increasingly rare in the past decade. Despite annual searches, this species has not been seen since 2006 and was the first North American bumble bee to be listed on the IUCN Red List of Threatened Species (critically endangered). The rapid decline of the Franklin Bumble Bee motivated researchers to start accumulating baseline data for the other North American species. Recent studies in both Canada and the United States have shown that many species are showing varying levels of declines; in fact, up to half of North American species may be at risk.

In 2012, the rare *B. affinis* was spotted in Madison, WI by a keen naturalist. CS

The Rusty-patched Bumble Bee (*B. affinis*) has the unfortunate honor of being the first federally listed endangered bee species in North America. As recently as the 1980s, it was one of the most common species throughout its large Canadian and US range; however, the Rusty-patched Bumble Bee has been found at only a few sites in recent years despite targeted searches by dozens of scientists and amateur naturalists. Recent sightings of this species have occurred in Minnesota, Wisconsin, and Illinois. There are other species in decline as well that

RIGHT: **Some evidence suggests that long-tongued bumble bees like *B. pensylvanicus* might be in decline.** TL

LEFT: **Certain species, such as *B. griseocollis*, remain common even in urban habitats.** SCO

have not yet been listed, such as various cuckoo species (*Psithyrus* spp.), the American Bumble Bee (*B. pensylvanicus*), and the Western Bumble Bee (*B. occidentalis*).

But not all North American bumble bee species are declining. Some species have become increasingly common, and some are even expanding their ranges naturally. The Common Eastern Bumble Bee (*B. impatiens*) and the Cryptic Bumble Bee (*B. cryptarum*) are rapidly becoming more common in regions where they were historically absent or rare. For example, the Common Eastern Bumble Bee is now found in parts of British Columbia, where it has been introduced for crop pollination. Some species, such as the Brown-belted Bumble Bee (*B. griseocollis*) and Vosnesensky Bumble Bee (*B. vosnesenskii*), are now particularly common in urban areas. Conservation management should therefore consider bumble bee species individually, rather than as a group.

THREATS TO BUMBLE BEES

The reason some bumble bee species are declining rapidly while others remain common is the subject of much scientific research. Bumble bee species differ in their seasonal activity, preferred food plants, colony productivity, habitat usage, and other life history traits. These differences may explain differential variability in regard to environmental stressors. According to research so far, it seems unlikely that one stressor is to blame in all situations, and a combination of threats may explain the declines. Below are some suspected threats to wild bumble bee populations.

— HABITAT LOSS

Many plants and animals have habitat requirements that are not met in our densely occupied cities and landscapes of intensified agriculture. Bumble bees need foraging habitat (e.g., fields of flowers blooming from spring until autumn), overwintering habitat (e.g., rotting logs and mulch), and nest habitat (e.g., abandoned rodent burrows and long grass), all in relative proximity to each other. Fragmentation is an issue: a study in Massachusetts found that railroads and highways partially restrict the movement of foraging bumble bees. In Iowa, bumble bee diversity and abundance has been shown to increase with floral resources in prairie remnants, and there is increasing evidence that bumble bee abundance can be promoted in our city parks, gardens, and green roofs with good design. The loss of fields, meadows, and forests filled with wildflowers and suitable nest and overwintering sites is thought to be one of the major causes of bee declines of many rarer species.

— INSECTICIDE USE

Farmers, golf-course managers, gardeners, and other landowners employ insecticides to deal with unwanted pest insects that may damage their flowering plants, crops, and lawns. Unfortunately, these insecticides almost inevitably have negative impacts on beneficial insects like bees. Bumble bees can be harmed by pesticides if they are sprayed on their nest site or on or near plants on which they are foraging. The insecticide can either be immediately toxic or have sublethal effects on foraging behavior and colony development over time. Particularly dangerous to bees are some systemic pesticides that are used to treat seeds and end up present throughout the plant tissues. These pesticides get into pollen and nectar, increasing the chance of contact by foraging bumble bees and their colonies. This problem of systemic pesticides is likely to be most important in agricultural areas.

— CLIMATE CHANGE

Bumble bees have evolved numerous physiological adaptations to particular climates and are generally better adapted to cooler climates than other bees. The emergence of queen bumble bees in the spring and the blooming of their food plants are both closely linked to changes in variables such as temperature and precipitation. An increased incidence of spring storms, drought, and other inclement weather patterns can cause the death of bumble bee queens and colonies. Bumble bees occurring in extreme environments at high elevations and in tundra ecosystems and those with narrow climatic tolerances might be particularly affected by climate change, but these effects have not yet been demonstrated for bumble bees.

— PATHOGEN SPILLOVER

In the early 1990s, managed bumble bees began to be used in North America to pollinate greenhouse and field crops such as tomatoes and cucumbers. Like many wild animals, bumble bees are more prone to disease while in captivity, and they may spread pathogens and parasites to wild bees when they escape from greenhouses. This phenomenon is known as *pathogen spillover*. When managed bees come in contact with native bees, the possibility of disease transfer increases. Wild bumble bees foraging near greenhouses with managed bumble bees have been found to have higher disease levels than wild bumble bees foraging far from greenhouses. The decline of some bumble bee species has been linked to the density of vegetable greenhouses in Canada and the United States. The severity of pathogen spillover from managed to wild bumble bees is not yet fully understood, but the phenomenon is thought to have contributed to the dramatic declines of *B. affinis* and other closely related species.

— INTRODUCTION OF EXOTIC AND INVASIVE SPECIES

The establishment of introduced bees, wasps, and plants alters the natural ecosystems within which native bumble bees have evolved. Some recently introduced bee species that may compete with bumble bees for resources include the Wool Carder Bee (*Anthidium manicatum*), the Giant Resin Bee (*Megachile sculpturalis*), and the Alfalfa Leafcutter Bee (*M. rotundata*). The European Honey Bee (*Apis mellifera*) was introduced to North America centuries ago and probably has little to do with recent declines of bumble bee species. However, high densities of feral and managed honey bee colonies remove available nectar and pollen resources, causing additional stress to declining bumble bee populations. A study in Colorado found that the abundance of honey bees can influence the abundance of short- and medium-tongued bumble bees in the summer. Additionally, successful native bumble bees themselves can compete with other native species. Bumble bee species that thrive in urban landscapes or are managed for agricultural crop pollination may establish in places where they naturally would not have occurred, thereby competing with the other native bumble bees for food and nest sites.

— INTENTIONAL AND ACCIDENTAL DEATHS

Because of their ability to sting, bumble bees are sometimes perceived to be dangerous and pest insects. Colonies are intentionally destroyed on urban and agricultural land out of fear that the bees will hurt humans and livestock. Additionally, bumble bees will be killed by passing cars if busy roads are present in their forage areas.

— HOW YOU CAN HELP

- Keep an eye out for rare species such as the Rusty-patched Bumble Bee (*Bombus affinis*) and Franklin Bumble Bee (*Bombus franklini*). If you see one, send photos to the Xerces Society (www.xerces.org/), Bumble Bee Watch (www.bumblebeewatch.org), or BugGuide (www.bugguide.net).
- Help document the presence of any bee species outside its native range by posting photos on sites such as BugGuide and Bumble Bee Watch.
- Provide food and shelter for bumble bees in gardens and fields by planting plants rich in pollen and nectar (see Forage Plant Guide).
- Support organizations undertaking pollinator conservation efforts in Canada and the United States such as Wildlife Preservation Canada, the Xerces Society for Invertebrate Conservation, and the Pollinator Partnership.
- Support organic agricultural production and gardening practices.
- Leave bumble bee colonies alone if you find any. If a colony has established somewhere on your property where you'd prefer them not to be, wait until winter to seal up the entrance, as the males and queens will have exited.

NATURAL ENEMIES

Like all insects, bumble bees face predators and parasites that constrain populations in natural ecosystems. Despite having a defensive stinger and warning coloration, bumble bees face numerous natural enemies. Bumble bee colonies are filled with protein- and carbohydrate-rich nectar, pollen, and larvae and are commonly attacked by mammals, including bears, raccoons, and skunks. In flight, a bumble bee may provide a quick snack for a bird.

Foraging bumble bees are regularly eaten by certain invertebrates. Crab spiders (Thomisidae) do not spin webs, but instead ambush bumble bees at flowers. Cryptically colored ambush bugs (Phymatidae) also attack bees on flowers. Robber flies (Asilidae), often close mimics of bumble bees, catch flying bees and inject them with enzymes before consuming their internal organs. Some wasps, including the Beewolf (Sphecidae; *Philanthus* sp.), capture live bees and paralyze them with specialized venom. The bees are then used to feed the wasp's developing larvae. Other predators include assassin bugs (Reduviidae) and dragonflies (Odonata).

Bumble bees are host to a variety of parasitoids, animals that lay their eggs inside their hosts' bodies, typically with the larvae developing inside the living host. Most bumble bee parasitoids are flies (Conopidae and Phoridae), and 30 percent or more of the workers in an area may be afflicted. Female flies attack hosts in flight, quickly inserting their ovipositor between abdominal terga to lay one or more eggs. After hatching, the larval fly feeds on host tissues and grows rapidly. Affected bees live for about two weeks before burying themselves in the ground and dying, after which the fly pupates and overwinters, emerging as an adult the following year. In addition to flies, North American bumble bees are host to at least one parasitoid wasp (Braconidae: *Syntretus* sp.).

A nematode worm, *Sphaerularia bombi*, enters overwintering queen bumble bees. The development of the worm causes sterilization of the young queen but does not necessarily reduce her life span. Afflicted queens can be observed foraging two to three weeks later than unaffected queens of the same species, by which time workers of the healthy colonies have taken over foraging.

Bumble bees are susceptible to numerous parasitic microorganisms that can have both lethal and sublethal effects. These pathogens can be transferred between bees within the colony, but they may also be picked up at flowers. Tracheal mites (*Locustacarus buchneri*) live in a bumble bee's air sacs and can cause decreased foraging efficiency. Protozoans and fungi (including *Apicystis bombi*, *Crithidia bombi*, and *Nosema bombi*) are commonly found in the bumble bee digestive

LEFT: **A *B. bimaculatus* male caught by a crab spider.** LR
BELOW: **Conopid flies are parasitoids of adult bumble bees.** LR

ABOVE: ***A B. impatiens* queen with external mites.** LR
RIGHT: **A microscopic view of *Nosema bombi*, a common gut parasite of bumble bees.** SR

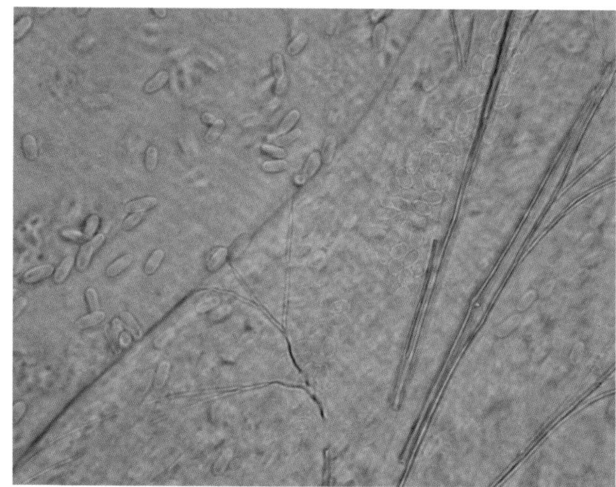

tract and fat bodies, where they consume gut contents or host tissue, with effects on bee life span, foraging ability, and overall colony fitness. Bumble bees are hosts to a variety of pathogenic viruses, fungi, and bacteria whose effects are poorly understood. Interestingly, these animals also harbor symbiotic bacteria that may protect them from pathogens. While bumble bees share long evolutionary histories with their parasites and pathogens, habitat loss, pesticide exposure, and genetic impoverishment may increase bees' vulnerability, and in some cases (e.g., the microscopic fungus *Nosema*) they have contributed to species declines.

MIMICRY

The eye-catching color patterns of bumble bees, mostly yellow and black, sometimes with red or white or both, serve to remind experienced predators that these bees may produce a painful sting when handled. Other insects, such as flower flies, which lack stingers or other distasteful characteristics, have evolved to look and act like bumble bees, thus gaining a measure of protection from predation through mimicry (see photo page 42) . In a classic case of such *Batesian mimicry*, the edible Viceroy butterfly has evolved to look like the distasteful Monarch, which sequesters cardiac glycosides when its caterpillars feed on milkweed.

The major groups that closely resemble bumble bees include those mentioned in the section titled Distinguishing Bumble Bees from Other Insects: flies, day-flying moths, beetles, sawflies, and other bees. Specific examples are listed below. Entering the genus and species name of these examples in a Web browser along with the word BugGuide will retrieve images of these fascinating mimics.

FLIES: Many flies, especially among the flower flies (Syrphidae; e.g., *Mallota posticata*) are good mimics of bumble bees. Others include robber flies (Asilidae; e.g., *Laphria astur*), bee flies (Bombyliidae; e.g., *Villa fulviana*), and deer bot flies (Oestridae; e.g., *Cephenemyia jellisoni*).

BEETLES: Among the beetles, the bumble flower beetle (Scarabaeidae; e.g., *Euphoria inda*) does not look very similar to bumble bees, but its flight behavior, including a loud buzzing sound when cruising low over vegetation, catches one's attention as a potential bumble bee until the brownish drab beetle alights on a flower. A beetle with greater color resemblance to bumble bees is the longhorned beetle (Cerambycidae; e.g., *Ulochaetes leoninus*).

LEFT: This syrphid fly is a very convincing bumble bee mimic. LR

BELOW: The locust borer (*Megacyllene robinae*) resembles bumble bees and is often found feeding on flowers. LR

These flower-feeding scarab beetles (*Trichiotinus* sp.) are thought to be bumble bee mimics. LR

DAY-FLYING MOTHS: Some clearwing hawk moths (Sphingidae; e.g., *Hemaris thetis*) look like large bumble bees as they fly through patches of flowers, but their slender legs and spindle-shaped antennae distinguish them immediately.
SAWFLIES: Some cimbicid sawflies (Cimbicidae; e.g., *Trichinosoma triagulum*) resemble bumble bees, but the swollen tips of their antennae distinguish them from bumble bees.
OTHER BEES: In addition to carpenter bees, mentioned in the section on distinguishing bumble bees from other insects, some digger bees (Apidae; e.g., *Anthophora bomboides*) bear a strong resemblance to bumble bees. The females, however, have dense broad brushes of hair on their hindlegs instead of a pollen basket for pollen transport, and males usually have a yellow clypeus.

Bumble bees themselves have carried mimicry a step further. In most areas where multiple species occur together, several of these often converge on one or two dominant color patterns. This *Müllerian mimicry*, in which several species possess a defensive sting, helps reinforce the avoidance training of potential predators for particular color patterns, thus reducing the loss of individuals from any one species.

Color convergence can make it difficult to distinguish different species of bumble bees in the field. It will often be necessary to capture and kill specimens for a microscopic examination of the more subtle diagnostic morphological features. With practice, capture for closer examination and release in the field may be enough, so that fewer bees need to be killed.

EXAMPLES OF MAJOR COLOR-PATTERN GROUPS

EASTERN NORTH AMERICA

PATTERN 1 Thorax mostly yellow with a dark central spot, abdomen yellow anteriorly: *B. impatiens, bimaculatus, griseocollis, affinis, vagans, sandersoni, perplexus,* and *fraternus.*
PATTERN 2 Thorax yellow anteriorly, mostly black posteriorly, abdomen with T2–3 yellow: *B. pensylvanicus, auricomus,* and *terricola.*

WESTERN NORTH AMERICA

PATTERN 3 (Pacific Coast): Thorax yellow anteriorly, black posteriorly, abdomen often with a yellow stripe: *B. vosnesenskii, caliginosus, vandykei,* some *fervidus, insularis, flavidus, crotchii, occidentalis,* and *franklini.*

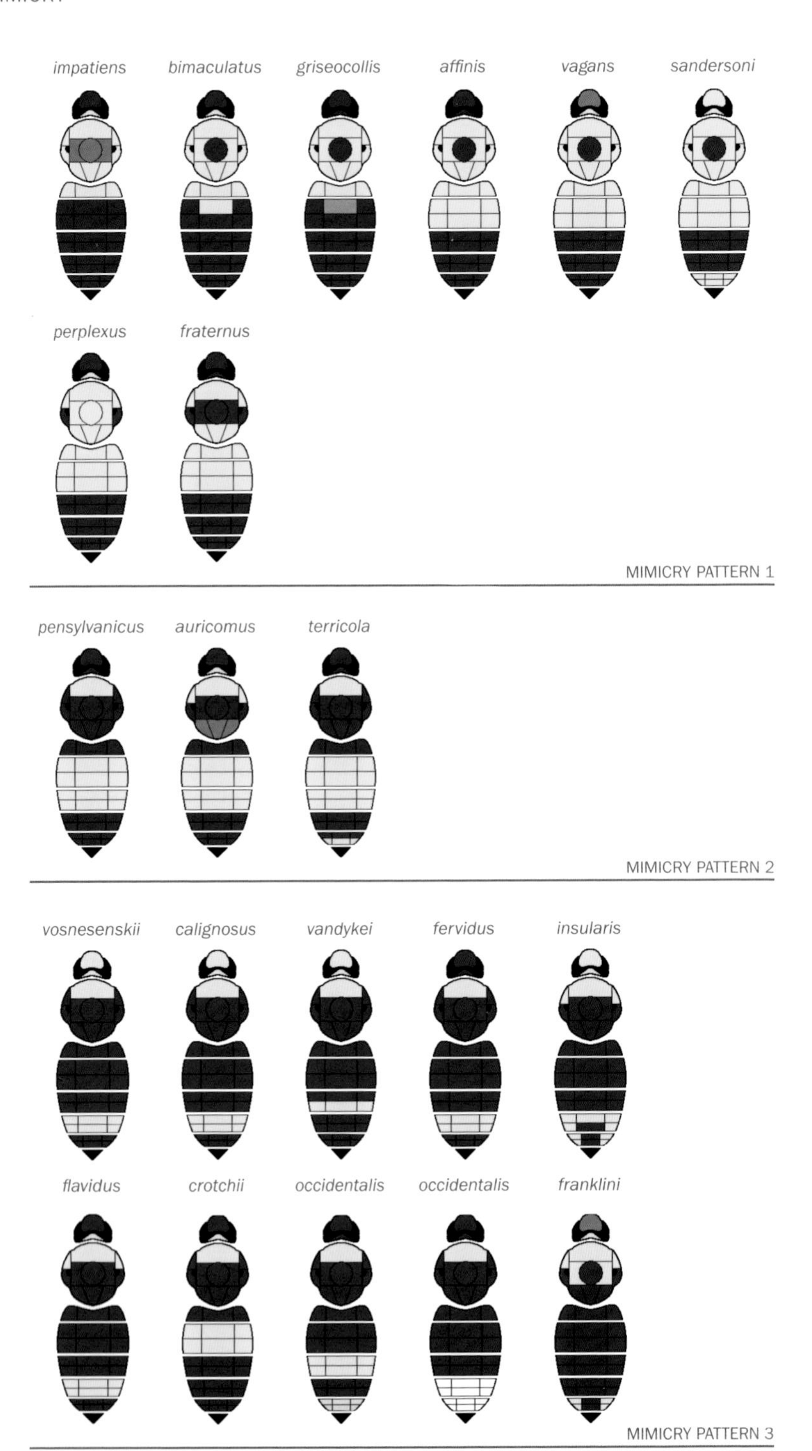
impatiens
bimaculatus
griseocollis
affinis
vagans
sandersoni
perplexus
fraternus
MIMICRY PATTERN 1
pensylvanicus
auricomus
terricola
MIMICRY PATTERN 2
vosnesenskii
calignosus
vandykei
fervidus
insularis
flavidus
crotchii
occidentalis
occidentalis
franklini
MIMICRY PATTERN 3

PATTERN 4 (California): Thorax yellow with black between wings, abdomen with T1 yellow, T2–3 black, T4 and sometimes T5 yellow or rusty: some *B. bifarius*, *melanopygus*, *sylvicola*, some *rufocinctus* and *mixtus*.

PATTERN 5 (Intermountain and Great Basin to southeastern Canada and northeastern USA): Thorax yellow with black between wings, abdomen with T2–3 red, T1 and T4 and sometimes T5 yellow: *B. huntii*, some *bifarius*, *sylvicola*, *melanopygus*, *ternarius*, and some *rufocinctus*.

Some widespread species are members of different Müllerian groups in different parts of their distribution ranges. *Bombus bifarius*, *melanopygus*, and *sylvicola* populations in most of California have black on T2–3, but these segments have red on them in populations to the north and east. *Bombus fervidus* is mostly yellow like *B. borealis* in the east, but becomes mostly black where it is found further west, resembling populations of several species of the Pacific Coast (Pattern 3).

The color patterns of males often generally resemble but are not always similar to those of females. The males of *B. vandykei*, usually all yellow, contrast starkly with the predominantly black females, but males are produced at the end of the colony season and are not as critical to perpetuation of the species as are the females and may be more expendable. This could help explain the lack of strong similarity of males to females. Males are not capable of stinging and might dilute the value of the warning coloration if discovered by predators.

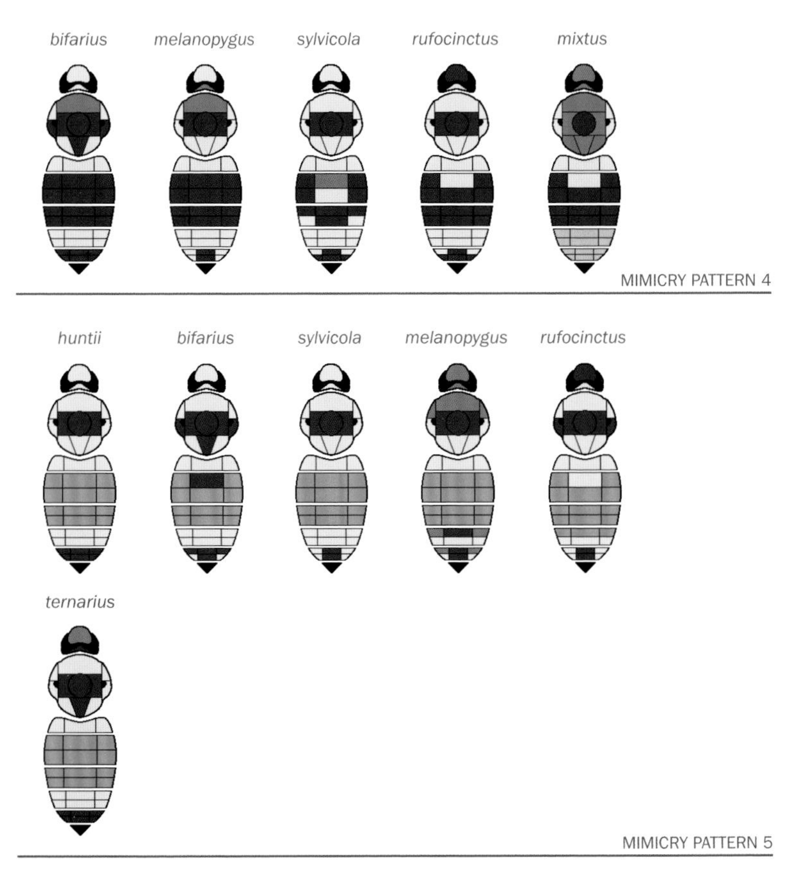

DISTINGUISHING BUMBLE BEES FROM OTHER INSECTS

Most people are familiar with bumble bees, recognizing their fuzzy, colorful, robust bodies and noisy bumbling flight between flowers. Bumble bees are very hairy bees with combinations of contrasting bright colors, mostly black and yellow, sometimes with various combinations of red or white. They have two pairs of wings that are usually folded back over the abdomen while they are foraging on flowers, or hooked together as a single unit when in flight. Bumble bees also have slender elbowed antennae, and females of the pollen-collecting species have the hind tibia expanded, slightly concave, and fringed with long hairs to form a pollen basket or *corbicula*. Bumble bees are unique among bees in lacking the jugal lobe of the hind wing, although the lack of this feature is visible only on close examination with a good hand lens or microscope.

Several groups of insects bear a general resemblance to bumble bees, including some robust flies, day-flying moths, beetles, and other Hymenoptera, especially certain sawflies and other bees. Characteristics of these other groups that will help distinguish them from bumble bees include the following:

FLIES have short stubby antennae and one pair of wings, and the compound eyes face more forward, even meeting at the top of the head in many male flies.

ABOVE: **The coloration of this syrphid fly (*bottom*) closely mimics *B. bifarius* (*top*), a co-occurring bumble bee.** AF

RIGHT: **This fly, found in the northeastern US, has a very similar color pattern to the widespread *B. ternarius*.** LR

Some bee flies Bombyliidae (left) and some hover flies Syrphidae (right) are frequently misidentified as bees because of their fuzzy bodies and nectar-feeding behavior. SCO

Some clearwing moths resemble bumble bees while foraging on flowers. LR

DAY-FLYING MOTHS have scales on their wings, long slender antennae and the tongue coiled beneath the head like a fire hose.

BEETLES have the front wings modified to form hard wing covers; their antennae are often long and slender or clubbed at the tips.

SAWFLIES have long slender antennae that may be swollen at the tips.

Leafcutter bees can be differentiated from bumble bees by the presence of pollen-carrying hairs on the underside of the abdomen instead of the legs. LR

OTHER BEES also have slender elbowed antennae, two pairs of membranous wings that hook together when in flight, and a narrow waist, but none except the introduced honey bee have the expanded hindleg pollen baskets like those of most female bumble bees.

People commonly confuse carpenter bees with bumble bees, but carpenter bees are usually much shinier and have less hair on top of their abdomens than bumble bees do. In contrast to bumble bees, the hind basitarsus of carpenter bees is longer than the tibia. Female carpenter bees also have a broad brush of hair on their hindlegs for pollen transport rather than the pollen basket found in bumble bees. At a microscopic level, carpenter bees have a very elongate (>7× as long as wide) marginal cell in the forewing.

The most challenging insects to distinguish from bumble bees are those that more specifically resemble bumble bees in color patterns and behavior, providing them a measure of protection from predators. This is discussed more fully in the section on Mimicry.

Carpenter bees (*Xylocopa* spp.) can be confused with bumble bees because of their large size and similar coloration. GZ (LEFT AND MIDDLE), LR (RIGHT)

BUMBLE BEE NAMES AND CLASSIFICATION

The aim of this guide is to help people identify bumble bees to species—that is, to find their correct names. Names are important because these labels, as part of the information retrieval system, allow us to bring together information from different sources on all aspects of each particular species, including their behavior and ecology.

For names to work for information retrieval, they have to be standardized. This is the purpose and advantage of formal "Latin" names. People have always given organisms "common" names, but these names are inconsistent in different languages and even in different places with the same language (e.g., "Yellow bumble bee" refers to a different species in Europe, where even in England the spelling of "bumblebee" is different). To address the problem of standardization, in the eighteenth century the Swede Carl Linnaeus developed a system now used throughout the world, with names usually based on classical Latin or Greek (which makes the names international and independent of any particular modern nation). For example, most children know the name *Tyrannosaurus rex*. Names in the Linnean system consist of a genus name (e.g., *Bombus* for all bumble bees), followed by a species name (e.g., *impatiens*), and often a reference to the original publication in which the name was first published (e.g., Cresson, 1863). More than one name may have been applied to the same species (synonyms), or names that look the same may have been applied to different species (homonyms). To keep the system in order, a set of rules is applied (see www.iczn.org/code).

Some groups of species share particular morphology or behavior. These differences are often predictable from the ancestral relationships among the species, because of the fundamental evolutionary model of descent with modification. Currently the best estimates of these relationships (in the sense that they have the most supporting evidence) often come from DNA-sequence data. These estimates can be represented as "family trees" of relationships among species. For bumble bees, the larger groups within these trees have been labeled with group names called subgenera (sing. subgenus), of which there are 15 worldwide.

Table 1 (pages 46–7) lists the 46 species of bumble bees known from America north of Mexico within their eight subgenera of *Bombus*. This list places the species most closely related by ancestry close together in the order, according to knowledge from morphology and from five genes. For example, bumble bees of the subgenus *Bombias* are the earliest to have diverged from the rest of the bumble bee species in North America. They have a peculiar nest architecture, with eggs laid in separate cells. North American species of the subgenera *Subterraneobombus* and *Thoracobombus* are related to similar grassland bees from Asia and Europe. These *Thoracobombus* bees are especially likely to nest among grass on the surface of the ground. Only two species are widespread in North America, but this group makes up a much larger part of the fauna of South America, especially at lower elevations. *Psithyrus* bees are the most distinctive group, because they are cuckoos in the nests of other bumble bees and lack the baskets (corbiculae) for transporting pollen on their hindlegs. The subgenus *Pyrobombus* is particularly diverse in North America (although absent from South America), perhaps in the niches occupied in the Old World by species of other subgenera. The North American species of *Pyrobombus* include the full range of desert, grassland, forest, mountain, and Arctic species. In contrast, *Alpinobombus* bees are striking for specializing in the tundra habitat of the very far north. Bees of the subgenus *Bombus s. str.*

TABLE 1 CHECKLIST OF THE 46 BUMBLE BEE SPECIES (GENUS *BOMBUS*) RECORDED FROM AMERICA NORTH OF MEXICO.

Species are grouped by subgenera based on estimated ancestral relationships (from evidence from five genes). Valid names of species are shown in bold with some of the more frequently encountered synonyms indented below them (marked "?" when synonymy is uncertain). In cases where a name was published originally within another genus, the reference is placed in parentheses and the original generic name shown in brackets. In some cases names cannot be used for these species because the same name was published earlier for another species (a homonym).

SUBGENUS	SPECIES
Bombias	***Bombus nevadensis*** Cresson, 1874
	Bombus auricomus (Robertson, 1903 [*Bombias*])
Subterraneobombus	***Bombus distinguendus*** Morawitz, 1869
	Bombus appositus Cresson, 1878
	Bombus borealis Kirby, 1837
Thoracobombus	***Bombus fervidus*** (Fabricius, 1798 [*Apis*])
	californicus Smith, 1854
	Bombus pensylvanicus (DeGeer, 1773 [*Apis*])
	americanorum Fabricius, 1804
	sonorous Say, 1837
Psithyrus	***Bombus variabilis*** (Cresson, 1872 [*Apathus*])
	?*intrudens* (Smith, 1861 [*Apathus*])
	Bombus citrinus (Smith, 1854 [*Apathus*])
	Bombus insularis (Smith, 1861 [*Apathus*])
	crawfordi (Franklin, 1913 [*Psithyrus*])
	Bombus suckleyi Greene, 1860
	Bombus bohemicus Seidl, 1837
	ashtoni (Cresson, 1864 [*Apathus*])
	Bombus flavidus Eversmann, 1852
	fernaldae (Franklin, 1911 [*Psithyrus*])
Pyrobombus	***Bombus vagans*** Smith, 1854
	bolsteri Franklin, 1913
	?*cockerelli* Franklin, 1913
	Bombus centralis Cresson, 1864
	Bombus flavifrons Cresson, 1863
	pleuralis Nylander, 1848
	dimidiatus Ashmead, 1902
	Bombus caliginosus (Frison, 1927 [*Bremus*])
	Bombus vandykei (Frison, 1927 [*Bremus*])
	cascadensis (Milliron, 1970 [*Pyrobombus*])
	Bombus melanopygus Nylander, 1848
	edwardsii Cresson, 1878
	Bombus sylvicola Kirby, 1837
	gelidus Cresson, 1878

SUBGENUS	SPECIES
Pyrobombus	***Bombus bimaculatus*** Cresson, 1863
	Bombus bifarius Cresson, 1878
	nearcticus Handlirsch, 1888
	Bombus ternarius Say, 1837
	Bombus huntii Greene, 1860
	Bombus vosnesenskii Radoszkowski, 1862
	Bombus impatiens Cresson, 1863
	Bombus perplexus Cresson, 1863
	Bombus sitkensis Nylander, 1848
	Bombus mixtus Cresson, 1878
	?*praticola* Kirby, 1837
	Bombus sandersoni Franklin, 1913
	Bombus frigidus Smith, 1854
	couperi Cresson, 1878
	Bombus jonellus (Kirby, 1802 [*Apis*])
	alboanalis Franklin, 1913
Alpinobombus	***Bombus polaris*** Curtis in Ross, 1835
	arcticus Kirby in Parry, 1824 (not of Quenzel, 1802)
	kincaidii Cockerell, 1898
	diabolicus Friese, 1911
	Bombus balteatus Dahlbom, 1832
	kirbiellus Curtis in Ross, 1835
	Bombus neoboreus Sladen, 1919
	strenuus Cresson, 1863 (not of Harris, 1776)
	Bombus hyperboreus Schönherr, 1809
	?*arcticus* (Quenzel in Acerbi, 1802 [*Apis*])
Bombus	***Bombus affinis*** Cresson, 1863
	Bombus franklini (Frison, 1921 [*Bremus*])
	Bombus occidentalis Greene, 1858
	mckayi Ashmead, 1902
	Bombus terricola Kirby, 1837
	Bombus cryptarum (Fabricius, 1775 [*Apis*])
	modestus Cresson, 1863 (not of Eversmann, 1852)
	moderatus Cresson, 1863
Cullumanobombus	***Bombus rufocinctus*** Cresson, 1863
	Bombus crotchii Cresson, 1878
	Bombus griseocollis (DeGeer, 1773 [*Apis*])
	Bombus morrisoni Cresson, 1878
	Bombus fraternus (Smith, 1854 [*Apathus*])

often have large colonies and are well known both as commercially reared pollinators of crops and as nectar robbers. (This subgenus bears the same name as the broader genus *Bombus* because it includes the "type species" of *Bombus*, *B. terrestris*; the abbreviation *s. str.* refers to the Latin term *sensu stricto*, meaning "in the strict sense," referring to the narrower sense of the name for the smaller group.) *Cullumanobombus* bees are predominantly grassland bees in North America, but more mountain specialists in South America.

As research progresses and new evidence becomes available, our understanding of the species and of their ancestral relationships changes. As a result, both the relationship trees and the names of the species sometimes need to change. This may appear to be a problem, but it is actually a strength, showing that we can update when we have improvements in our understanding. In Table 1, following the current valid name for each species (in bold) is a list of synonyms. These are names for bumble bees previously considered separate species, but now thought to be members of the same species. This process of discovery continues with this guide, which applies changes to the names for a few species, because new evidence from DNA barcodes (from the COI gene) provides stronger support for interpretations of their status as parts of species more widespread elsewhere in the world.

HOW TO USE THIS BOOK TO IDENTIFY BUMBLE BEE SPECIES

The hair (pile or pubescence) of bumble bees has many different color patterns, which may give the impression that species should be easy to identify. Unfortunately it is not that straightforward. Simple keys based on color patterns may appear easy to use, and may work well on small local faunas, but they are unreliable for correct identification at the continent-wide level. Not only do bumble bee color patterns often vary a lot within species, but different species can also look very similar to one another. It is also possible that new color variations will always be found. So even after more than 200 years of study, some species remain difficult to identify. This difficulty may be unfortunate, but it challenges people to find new and more effective diagnostic characters for identifying the species.

This guide differs from most previous identification aids in recognizing that a lot of color-pattern variation exists within many bumble bee species, and it emphasizes the importance of this reality for successful identification. Multiple color-pattern diagrams are presented here for each species in order to describe the variation and to facilitate comparisons. Inevitably these diagrams are simplifications and compromises. Our diagrammatic scheme for color patterns cannot show the most subtle variations; it is designed to provide a user-friendly guide to the major differences. These diagrams make it easier to see precisely which body segment has hair of which color, something that may not be easy to see from a photograph. The body surface underneath is always black or brown (color patterns described here refer only to the hairs). So for these diagrams, relatively few body regions are distinguished, and these are classified into a limited range of colors, adopting the approximate color that covers most or is most apparent within each body region (e.g., small pale spots will not be precisely represented). A nearly equal mixture of black and yellow hairs is shown as olive, and a nearly equal mixture of black and near-white hairs is shown as gray. Each diagram represents a single real bumble bee examined and no doubt others will be found with different color patterns.

For best results, start with the main keys for females and males (pages 168–98). Females are often the majority of individuals. They have 6 segments (terga) visible dorsally on the metasoma (rather than 7 for males), the tip of their abdomen is usually more pointed, their antennae are short with 12 segments (rather than longer, with 13 segments for males), and their mandibles do not have a beard of dense long hairs. Using these keys is hard work even for specialists, but they are more reliable than using color pattern alone. Ideally, these keys should be applied to at least a few reference specimens from any one place. Then other specimens can be compared with them.

To achieve more reliable species identifications, the main keys use not only the color patterns, but also the more difficult morphological characters. Morphological characters may be clearly visible only under a microscope (or with a powerful hand lens), but crucially they are much less variable within species than color. For examination with a microscope, specimens will usually need to have been collected and placed on a long pin, so they can be handled easily without being broken. The hair needs to be in fresh, unmatted condition. Matted dead bees can be cleaned by immersing them in warm water with a small amount of detergent, then rinsing them in clean water. A final rinse in alcohol can help to dry them with the hair erect. The hair can even be fluffed up with a fine soft brush, such as a camera-lens brush. It is useful to extend the legs and mandibles with forceps, although this may not be easy. Male bees must have their genitalia (a small hard capsule) visible for examination. The male genitalia of fresh specimens can be hooked out with a pin and exposed from between the last visible tergum and sternum of the metasoma (the *metasoma* is the abdomen, but only from segment 2 onward, after the bee's narrow waist; the two are different). It is better to leave the genitalia attached to the specimen if possible, so they do not become separated and lost, otherwise they can be glued to a card and pinned with the bee. Unfortunately, diagnostic morphological characters are not always easy to see or use, and sometimes not as many are known as we might wish. When unfamiliar special terms for morphological structures are used, this is only because they are needed in order to label structures with precision.

Alternatively, use the quick key below to place a specimen in one of the four principal groups used in this guide. Although also not always easy, with growing experience this quick key helps separate many similar-looking species, greatly reducing misidentifications at subsequent steps. Then check in the guide within the appropriate group (these are color-coded to help navigation) against the color-pattern diagrams. Within each of the groups, species are arranged to place those with the most similar color patterns together, to help in making comparisons of the differences. Finally, check the specimen against the diagnostic characters given in the identification notes for each species. At the top of the species notes are lists of the most similar-looking species (with species in other groups given in parentheses). Within the species notes, after each diagnostic character, the shorthand "contrast" is used to point out the most similar species that have differences in each particular diagnostic character. Check the notes on these species to confirm the differences. The most important diagnostic characters are shown in italics.

The quick key below can be used to assign female bumble bees (queens and workers) to one of the four principal groups in this guide. Keys are arranged in pairs of paragraphs ("couplets") that describe a series of contrasting states of diagnostic characters. Choose the paragraph that agrees most closely with a particular specimen and then move to the next numbered couplet or species as indicated at the end of the paragraph. Usually the first character described in a key couplet will be the most reliable, even if unfortunately it is not always the easiest to see.

In this guide, the length of the cheek (technically known as the "oculo-malar area") is measured as the minimum distance between (on the upperside) the compound eye lower edge and (on the lowerside) the cheek edge between the mandibular hinges. The cheek breadth is measured as the distance between and including the mandibular hinges.

QUICK KEY FOR FEMALES TO THE FOUR GROUPS

1a midleg (*basitarsus* back far) corner rounded (>45°) → **2**

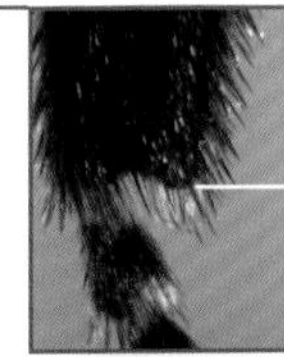

1b midleg corner with a sharp spine (<45°) → **3**

2a (1a) cheek longer than broad or as long as broad ("square") → **Group 1** *(page 52)*

2b cheek shorter than broad → **Group 2** *(page 111)*

3a (1b) hindleg (tibia outer surface) flat without long hair (right) → **Group 3** *(page 136)*

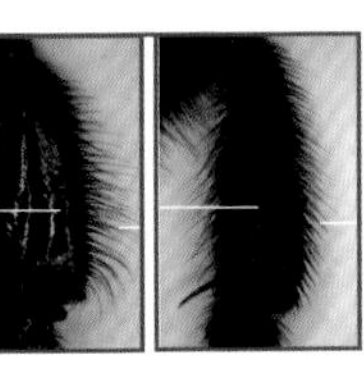

3b hindleg convex with long hair (far right) → **Group 4** *(page 155)*

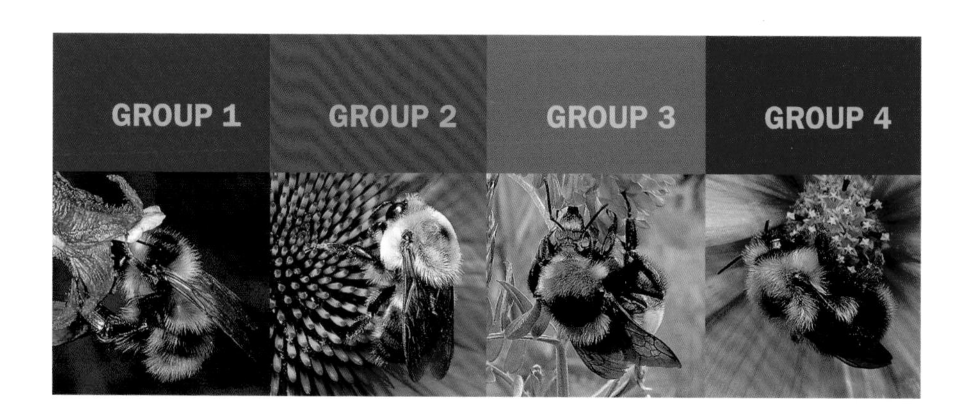

SPECIES ACCOUNTS

BOMBUS VOSNESENSKII RADOSZKOWSKI, 1862
VOSNESENSKY BUMBLE BEE

Worker *Bombus vosnesenskii*. GZ

Male *Bombus vosnesenskii*. DK

IDENTIFICATION

Western, medium tongue-length species. Most similar to *B. caliginosus* and *B. vandykei* (see also *B. occidentalis, B. franklini, B. fervidus, B. insularis,* and *B. flavidus*).

HAND CHARACTERS Body size medium (larger than *B. caliginosus, B. vandykei*): queen 18–21 mm (0.69–0.83 inch), worker 8–17 mm (0.33–0.65 inch). Hair short and even. Head length medium with the cheek (oculo-malar area) as long as broad (contrast *B. occidentalis, B. franklini, B. fervidus*), midleg basitarsus with the back far corner rounded, hindleg tibia outer surface flat without long hair but with long fringes at the sides, forming a pollen basket (corbicula). Hair of metasomal *T3 black* (contrast *B. vandykei*), T4 almost entirely yellow with just a few black hairs near the midline (contrast *B. caliginosus, B. vandykei*), *S2–5 with black fringes at the back* (contrast *B. caliginosus*) or very rarely with yellow at the sides. Male 10–15 mm (0.40–0.58 inch). Eye similar in size and shape to the eye of any female bumble bee. Antenna of medium length, flagellum 3× longer than the scape (contrast *B. occidentalis, B. franklini, B. fervidus*). Hair color pattern similar to the queen/worker, but metasomal T5 at the sides with yellow, rarely with yellow intermixed broadly on the upperside of the thorax at the back.

MICROSCOPIC CHARACTERS Queen/worker lower central area of the clypeus with many large pits or punctures (contrast *B. caliginosus*). Male genitalia with the penis-valve head sickle-shaped, the back-curved "sickle" long and very narrow, at least 3× longer than its breadth in the further half and less than half the breadth of the adjacent neck of the penis-valve head, the "sickle" scarcely flattened and about 2× broader than thick, almost spinelike (contrast *B. caliginosus, B. vandykei*), gonostylus longer than broad (contrast *B. caliginosus, B. vandykei*), the inner (medial) edge concave.

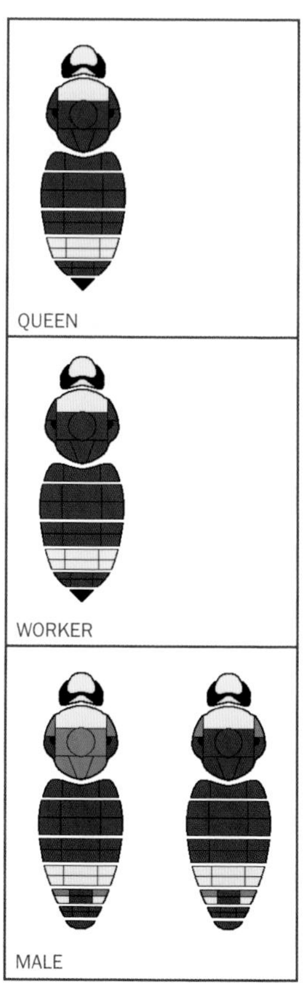

OCCURRENCE

RANGE AND STATUS Mediterranean CA and Mountain West of CA, OR, WA, southern BC and adjacent Desert West in CA. From sea level to above 2700 m. One of the most common species near the west coast.

HABITAT Open grassy areas, urban parks and gardens, chaparral and shrub areas, mountain meadows.

EXAMPLE FOOD PLANTS *Arctostaphylos, Ceanothus, Chrysothamnus, Cirsium, Eriogonum, Eschscholzia, Lupinus, Phacelia, Rhododendron, Ribes, Vicia, Ericameria, Clarkia, Grindelia.*

BEHAVIOR Nests underground. Males patrol circuits in search of mates.

PARASITISM BY OTHER BEES Unknown.

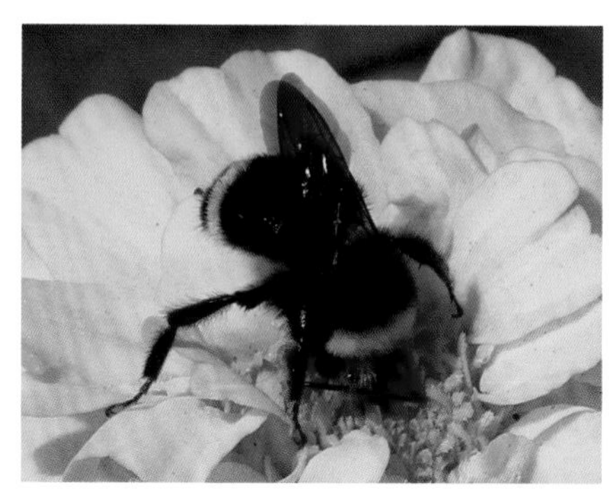

Male *Bombus vosnesenskii*. DK

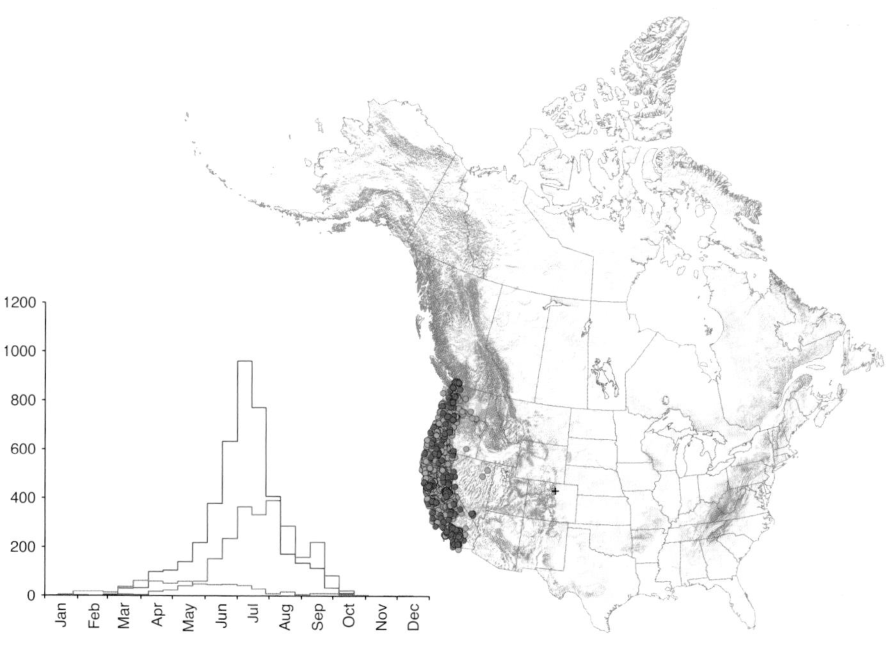

BOMBUS CALIGINOSUS (FRISON, 1927)
OBSCURE BUMBLE BEE

ABOVE: **Female *Bombus caliginosus*.** AR
INSET: **Male (posed) *Bombus caliginosus*.** SCO

IDENTIFICATION

Western coastal, medium long-tongued species. Most similar to *B. vosnesenskii* and *B. vandykei* (see also *B. occidentalis, B. franklini, B. fervidus, B. insularis,* and *B. flavidus*).

HAND CHARACTERS Body size small (smaller than *B. vosnesenskii,* larger than *B. vandykei*): queen 16–18 mm (0.62–0.71 inch), worker 11–13 mm (0.43–0.50 inch). Hair short but slightly shaggy and uneven. Head length medium with the cheek (oculo-malar area) very slightly longer than broad (contrast *B. occidentalis, B. franklini, B. fervidus*), midleg basitarsus with the back far corner rounded, hindleg tibia outer surface flat without long hair but with long fringes at the sides, forming a pollen basket (corbicula). Hair of metasomal *T3 black* (contrast *B. vandykei*), T4 yellow

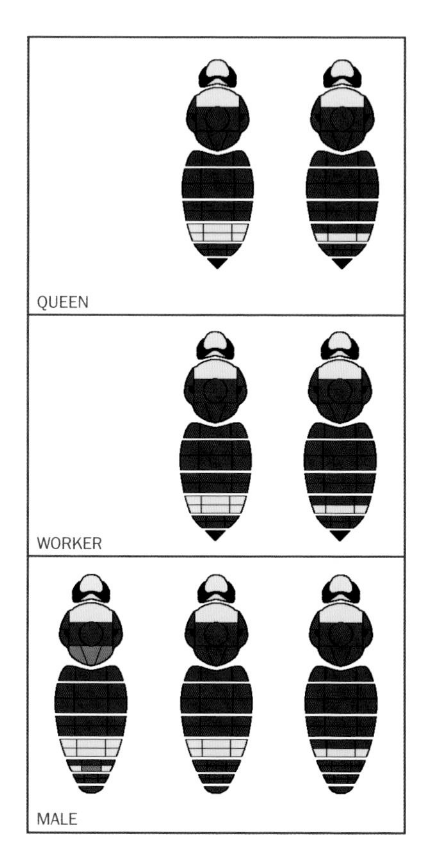

but with black at the front and often forming a triangle in the middle and narrowly interrupting the yellow along the midline (contrast *B. vosnesenskii, B. vandykei*), *S4 or S3–4 usually with pale fringes at the sides at the back* (contrast *B. vosnesenskii*) or very rarely nearly entirely black, S2 and S5 with black fringes. Male 11–13 mm (0.46–0.49 inch). Eye similar in size and shape to the eye of any female bumble bee. Antenna of medium length, flagellum 3× longer than the scape (contrast *B. occidentalis, B. franklini, B. fervidus*). Hair color pattern similar to the queen/worker.

MICROSCOPIC CHARACTERS Queen/worker lower central area of the clypeus with only small and not large pits or punctures (contrast *B. vosnesenskii*). Male hair of the upperside of the thorax at the back (scutellum) with a few yellow hairs intermixed, metasomal T4 with some black hair at the front, T5 yellow at the sides and sometimes at the back. Antennal segment A3 long, length nearly 2× the maximum breadth, almost as long as antennal segment A5. Genitalia with the penis-valve head

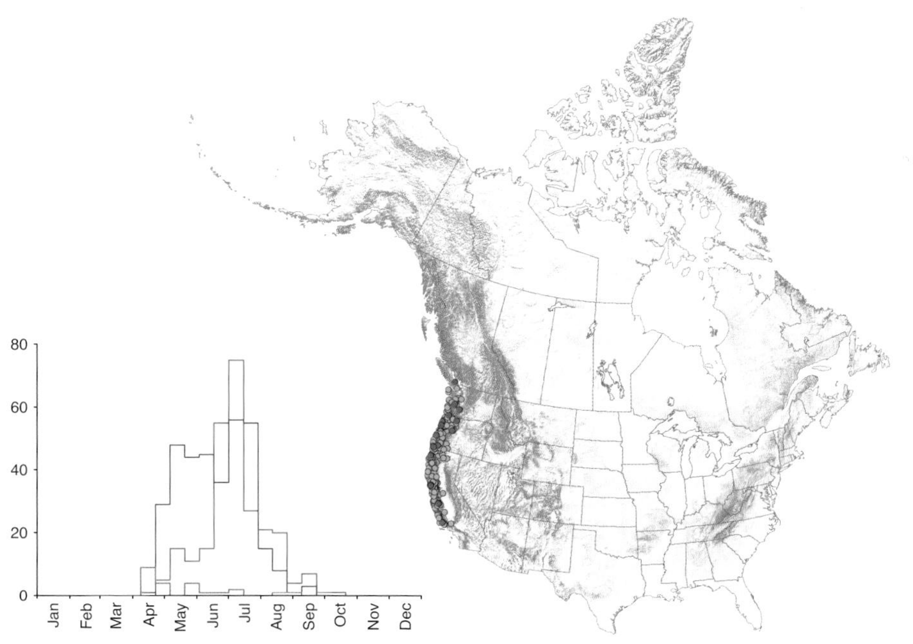

Male *Bombus caliginosus*. VL

sickle-shaped, *the back-curved "sickle" short and broad* (contrast *B. vosnesenskii*) less than 2× longer than the breadth of the broadest part and similar in breadth to the adjacent neck of the penis-valve head, the "sickle" flattened and broadly rounded at the tip, gonostylus shorter than broad (contrast *B. vosnesenskii*), the inner (medial) edge nearly straight, the margin thin with a short indistinct parallel submarginal groove.

OCCURRENCE

RANGE AND STATUS Mediterranean CA and Pacific Coast from southern CA to southern BC, with scattered records from the east side of CA Central Valley. Uncommon.

HABITAT Open grassy coastal prairies and Coast Range meadows.

EXAMPLE FOOD PLANTS *Ceanothus, Cirsium, Clarkia, Keckiella, Lathyrus, Lotus, Lupinus, Rhododendron, Rubus, Trifolium, Vaccinium.*

BEHAVIOR Nests underground, also nests aboveground in abandoned bird nests. Males patrol circuits in search of mates.

PARASITISM BY OTHER BEES Unknown.

BOMBUS VANDYKEI (FRISON, 1927)
VAN DYKE BUMBLE BEE (INCLUDING *CASCADENSIS*)

ABOVE: **Female *Bombus vandykei*.** GZ
RIGHT: **Male *Bombus vandykei*.** HW

IDENTIFICATION

Western, medium long-tongued species. Most similar to *B. vosnesenskii* and *B. caliginosus* (see also *B. occidentalis, B. franklini, B. fervidus, B. insularis,* and *B. flavidus*). The extensively yellow female color pattern (unnecessarily redescribed with the name *cascadensis,* from WA, northern OR) is rare in collections compared to the darker female color patterns from southern OR, CA.

HAND CHARACTERS Body size small (smaller than *B. vosnesenskii, B. caliginosus*): queen 14–18 mm (0.56–0.72 inch), worker 10–14 mm (0.39–0.50 inch). Hair short but slightly shaggy and uneven. Head length medium with the cheek (oculo-malar area) longer than broad (contrast *B. occidentalis, B. franklini*), midleg basitarsus with the back far corner rounded, hindleg tibia outer surface flat without long hair but with long fringes at the sides, forming a pollen basket (corbicula). Hair of metasomal *T3 yellow at least in the back half* (contrast *B. vosnesenskii, B. caliginosus*) and usually with some black at the front and in the middle, T4 usually black but T1–2 and T4 may rarely have small yellow patches at the sides, S2–5 with pale fringes at the back. Male 10–13 mm (0.41–0.52 inch). Eye similar in size and shape to the eye of any female bumble bee. Antenna of medium length, flagellum 2.5–3× longer than the scape (contrast *B. occidentalis, B. franklini, B. fervidus*). Hair usually *predominantly yellow* (contrast *B. vosnesenskii, B. caliginosus*), upperside of the thorax between the wings with black intermixed in a narrow band, metasomal T3–4 usually yellow, without black intermixed, but rare dark males have yellow only on the upperside of the thorax at the front and on T2–3 with black behind the wings and on T1 (contrast some *B. flavifrons*).

MICROSCOPIC CHARACTERS Male antennal segment A3 long, length nearly 2× maximum breadth, almost as long as antennal segment A5. Genitalia with the penis valve sickle-shaped, *the back-curved "sickle" short and broad* (contrast *B. vosnesenskii*), about 2× longer than the breadth of the broadest part and similar in breadth to adjacent neck of the penis-valve head, the "sickle" flattened and broadly rounded at the tip, gonostylus shorter than broad (contrast *B. vosnesenskii*), the inner (medial) edge nearly straight, the margin thin with a short indistinct parallel submarginal groove.

OCCURRENCE

RANGE AND STATUS Mediterranean CA and Mountain West of CA, OR, WA, to southern BC. From the Transverse Ranges, mostly through the Sierra-Cascade Ranges and sparingly through the Coast Ranges. From sea level up to 2,200 m. Moderately common.

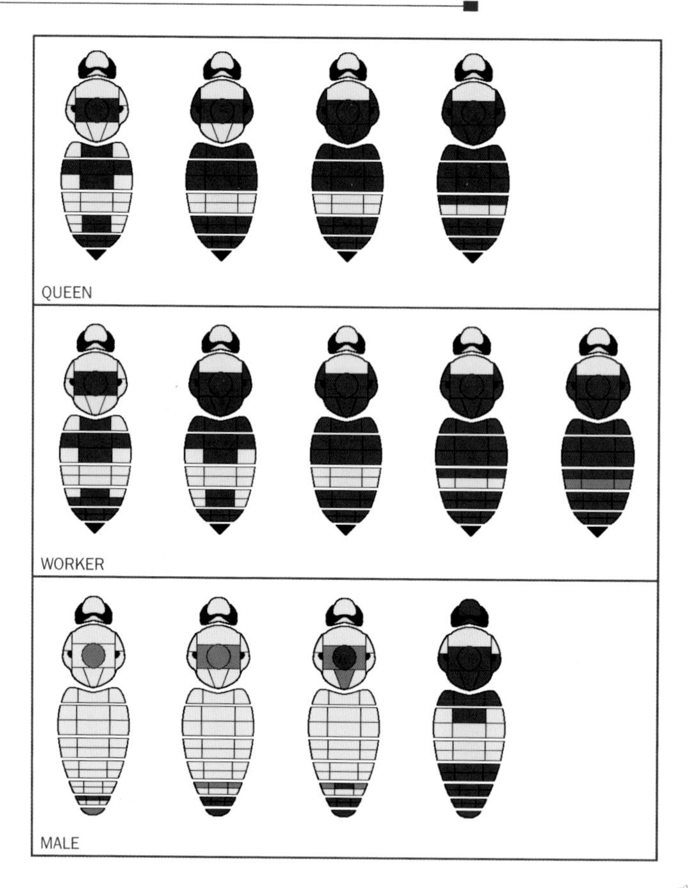
QUEEN
WORKER
MALE

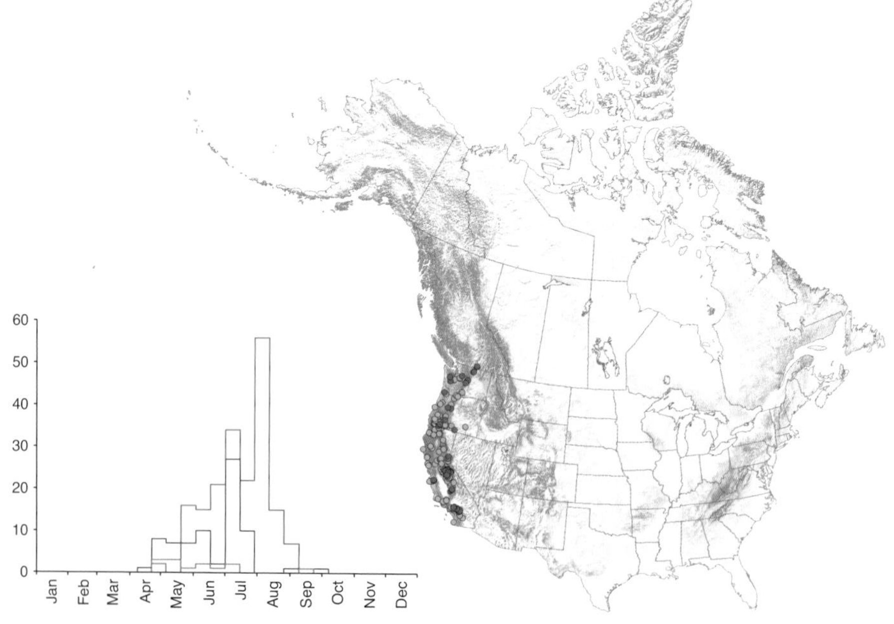
60
50
40
30
20
10
0
Jan
Feb
Mar
Apr
May
Jun
Jul
Aug
Sep
Oct
Nov
Dec

HABITAT Open grassy prairies and meadows.
EXAMPLE FOOD PLANTS *Cirsium, Clarkia, Collinsia, Delphinium, Eriodyction, Lupinus, Penstemon, Phacelia, Salvia, Stachys, Trifolium.*
BEHAVIOR Nests underground. Males patrol circuits in search of mates.
PARASITISM BY OTHER BEES Unknown.

ABOVE: **Queen *Bombus impatiens.*** LR
ABOVE RIGHT: **Male *Bombus impatiens.*** LR
RIGHT: **Worker *Bombus impatiens.*** LR

IDENTIFICATION

Eastern, medium-tongued species. Most similar to *B. bimaculatus, B. perplexus, B. vagans,* and *B. sandersoni* (see also *B. citrinus, B. variabilis, B. griseocollis,* and some *B. rufocinctus*).
HAND CHARACTERS Body size large: queen 21–23 mm (0.81–0.90 inch), worker 9–14 mm (0.34–0.56 inch). Hair short and even. Head length medium with the cheek (oculo-malar area) as long as broad or just shorter than broad (contrast *B. griseocollis, B. rufocinctus*), midleg basitarsus with the back far corner rounded, hindleg tibia outer surface flat without long hair but with long fringes

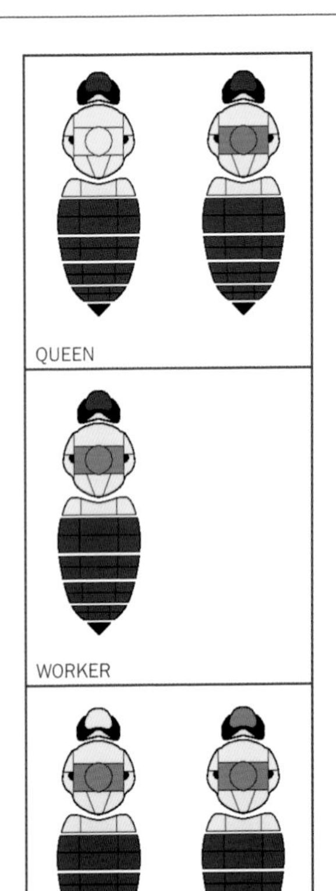

Worker *Bombus impatiens*. LR

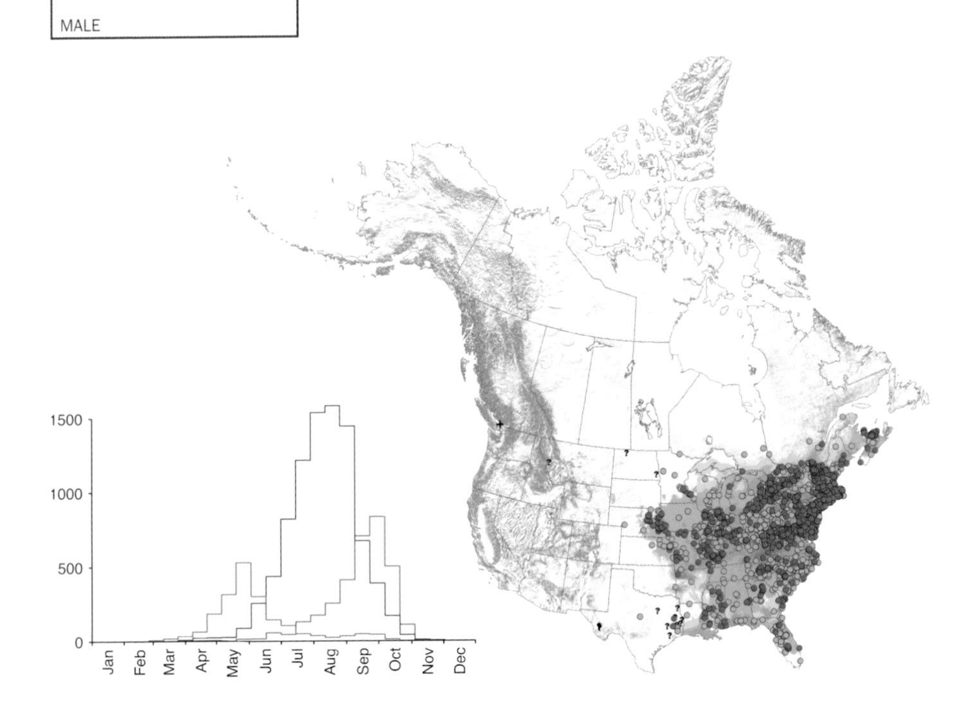

at the sides, forming a pollen basket (corbicula). Hair of the face black or with a few yellow hairs intermixed, thorax predominantly yellow, the upperside with short black hairs intermixed diffusely in a large square between the wings, metasomal T1 yellow, *T2 usually entirely black* (contrast *B. bimaculatus, B. perplexus, B. vagans, B. sandersoni*), tail black. Rarely T2–3 with orange-red hair at the back, although this may be from damage at the pupal stage. Very rarely worker T2 with a few yellow hairs at the front near the middle. Body rather long and rectangular. Male 12–14 mm (0.48–0.54 inch). Eye similar in size and shape to the eye of any female bumble bee (contrast *B. griseocollis, B. rufocinctus*). Antenna of medium length, flagellum 3× longer than the scape. Hair color pattern similar to the queen/worker, but below the antenna a patch of yellow, and on the underside and on the legs and metasomal S2–6 many yellow hairs. Rarely T2 with a few yellow hairs at the front and in the middle, very rarely T4–5 with small patches of yellow at the sides.

MICROSCOPIC CHARACTERS Queen metasomal T6 near the tip flat or with a weak lengthwise ridge just before the tip (contrast *B. perplexus*). Male genitalia with the penis valve sickle-shaped, the back-curved "sickle" long and narrow, at least 3× longer than broad and less than half the breadth of the adjacent neck of the penis-valve head, the "sickle" scarcely flattened and about 2× broader than thick, almost spinelike, gonostylus with the inner (medial) edge concave.

OCCURRENCE

RANGE AND STATUS One of the most widespread and abundant species in the Eastern Temperate Forest region of the eastern US and adjacent southern Canada, as far west as the eastern Great Plains. Possibly expanding in range toward the northeast. Commonly used for pollination of greenhouse crops. Apparently naturalized from escapes in southern BC and perhaps elsewhere.

HABITAT Close to or within woodland, grassland, farmland, wetlands, urban parks and gardens.

EXAMPLE FOOD PLANTS *"Aster", Cirsium, Eupatorium, Gelsemium, Impatiens, Malus, Pontederia, Rubus, Solidago, Trifolium.*

BEHAVIOR Nests underground. Males patrol circuits in search of mates.

PARASITISM BY OTHER BEES Host to *B. citrinus*, confirmed breeding record.

BOMBUS BIMACULATUS CRESSON, 1863
TWO-SPOTTED BUMBLE BEE

ABOVE: **Female *Bombus bimaculatus*.** LR
ABOVE RIGHT: **Male *Bombus bimaculatus*.** LR
RIGHT: **Worker *Bombus bimaculatus*.** SCO

IDENTIFICATION

Eastern, medium-tongued species. Most similar to *B. impatiens, B. perplexus, B. vagans,* and *B. sandersoni* (see also *B. griseocollis,* some *B. rufocinctus, B. affinis, B. citrinus,* and *B. variabilis*). The lightest and darkest female color patterns are rare.

HAND CHARACTERS Body size medium: queen 18–22 mm (0.71–0.85 inch), worker 10–16 mm (0.38–0.63 inch). Hair long and uneven (contrast *B. impatiens, B. griseocollis*). Head length medium with the cheek (oculo-malar area) just longer than broad (contrast *B. griseocollis, B. rufocinctus*), midleg basitarsus with the back far corner rounded, hindleg tibia outer surface flat without long hair but with long fringes at the sides, forming a pollen basket (corbicula). Hair of the face black or with only a few yellow hairs intermixed, upperside of thorax between the wings with a dense black spot always present although often with yellow hairs intermixed, only occasionally forming a black band between the wings, sides of the thorax yellow (contrast *B. perplexus*), metasomal *T2 usually extensively black at least at the sides in front* (contrast *B. vagans, B. sandersoni*), *always with a small patch of yellow present in the front near the midline* (contrast *B. impatiens*) but this varies in extent, although the back edge of the yellow usually *forming a characteristic W shape* (which may be similar in a few *B. griseocollis,* which in contrast have very short hair), or if the hair of T2 is extensively yellow, then it at least has many black hairs intermixed in the back corners at the sides (contrast *B. impatiens, B. perplexus, B. vagans, B. sandersoni*). Metasoma rather short and globular. Male 12–15 mm (0.46–0.59 inch). Eye similar in size and shape to the eye of any female bumble bee (contrast *B. griseocollis, B. rufocinctus*). Antenna of medium length, flagellum 3× longer than the scape. Hair color pattern similar to the queen/worker, but

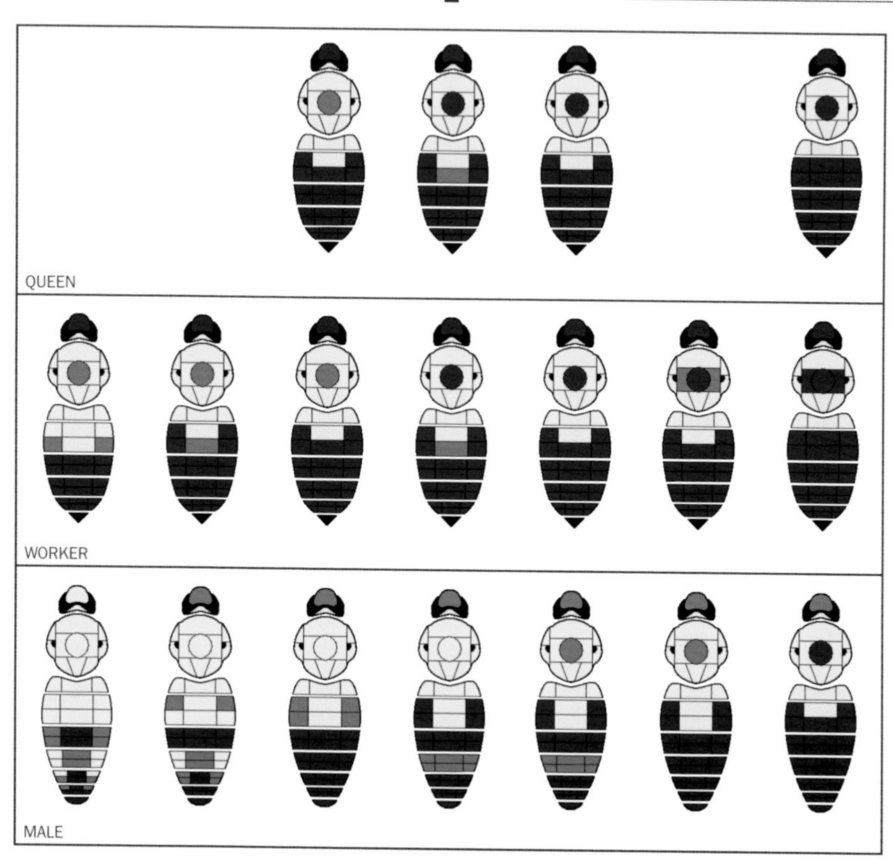
QUEEN
WORKER
MALE

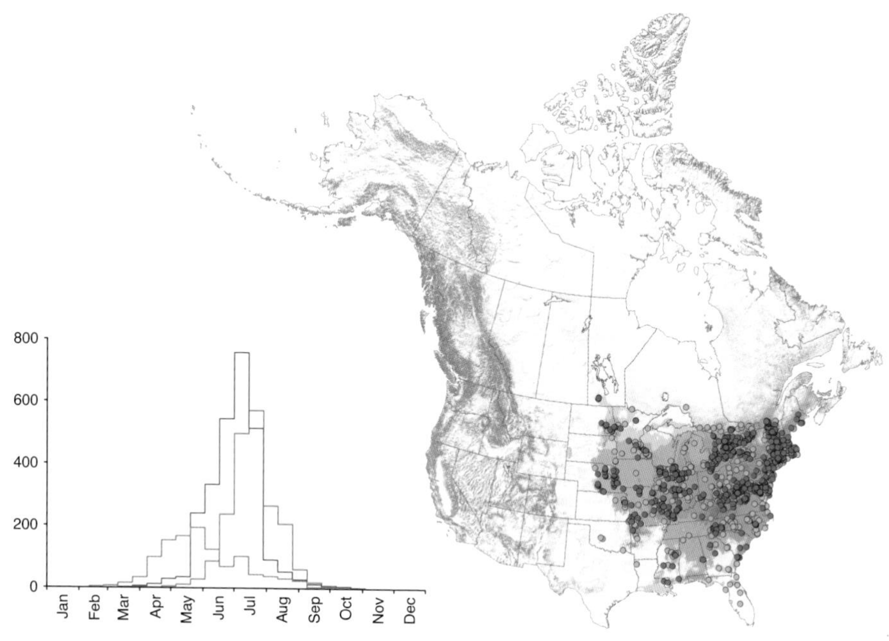
800
600
400
200
0
Jan
Feb
Mar
Apr
May
Jun
Jul
Aug
Sep
Oct
Nov
Dec

below the antenna a patch of yellow intermixed with black, metasomal T4–5 at the sides often with at least some yellow hair.

MICROSCOPIC CHARACTERS Queen metasomal T6 flat or with a weak lengthwise ridge just before the tip (contrast *B. perplexus*). Male genitalia with the penis valve sickle-shaped, at the far end back-curved as a long and narrow "sickle" at least 3× longer than broad, but distinctly flattened and more than 3× broader than thick, rounded and not expanded at the tip, the penis-valve angle on the underside of the shaft and to the side only slightly closer to the base than to the head of the penis valve, gonostylus with the inner (medial) edge concave.

OCCURRENCE

RANGE AND STATUS One of the most widespread and abundant species in the eastern US and adjacent southern Canada, throughout the Eastern Temperate Forest region, although less common on the southeastern US coastal plain, in the eastern Boreal Forest, and on the eastern Great Plains.

HABITAT Close to or within wooded areas, urban parks and gardens.

EXAMPLE FOOD PLANTS *Campanula, Lonicera, Monarda, Prunus, Rhododendron, Rosa, Rubus, Tilia, Trifolium, Vaccinium, Vicia.*

BEHAVIOR Nests mostly underground, though occasionally aboveground or in cavities. Males patrol circuits in search of mates. One of the species with colonies that end earliest in the summer.

PARASITISM BY OTHER BEES Host to *B. citrinus*, confirmed breeding record.

BOMBUS PERPLEXUS CRESSON, 1863
CONFUSING BUMBLE BEE

Female *Bombus perplexus*. LR

Male *Bombus perplexus*. SCO

IDENTIFICATION

Eastern and northern, medium-tongued species. Most similar to *B. impatiens, B. bimaculatus, B. vagans,* and *B. sandersoni* (see also *B. griseocollis, B. rufocinctus,* and *B. affinis*). Evidence from DNA barcodes supports a close relationship between *B. perplexus* and taxa of the *B. hypnorum*-complex from northern Europe and Asia, although it appears to be a separate species.

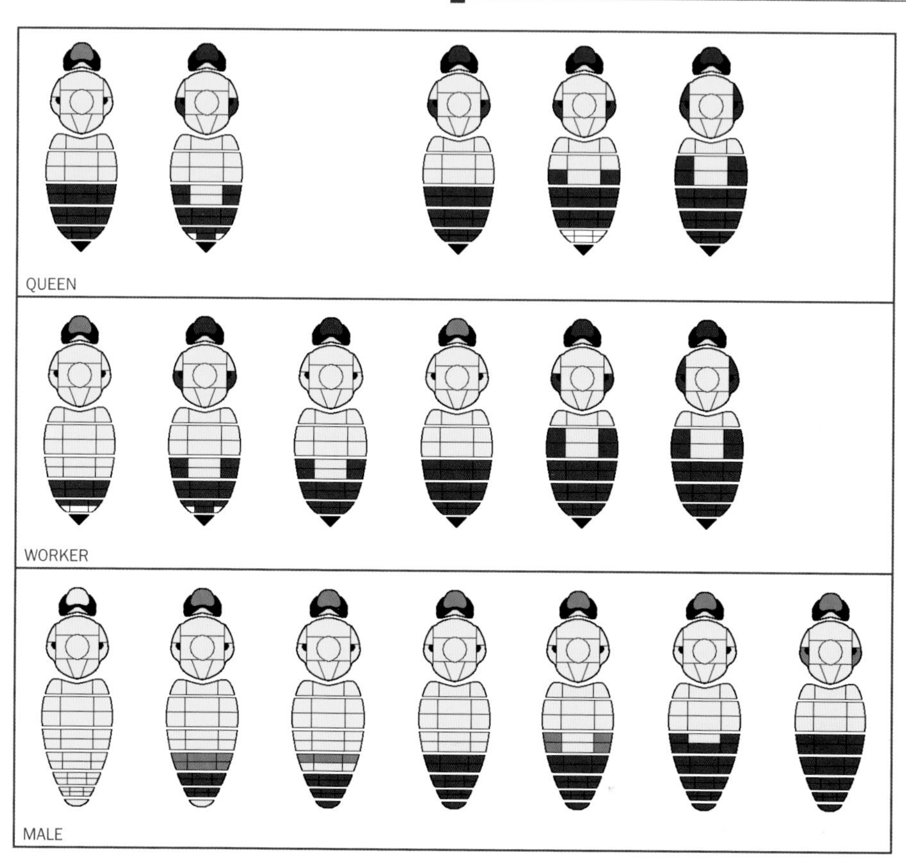
QUEEN
WORKER
MALE

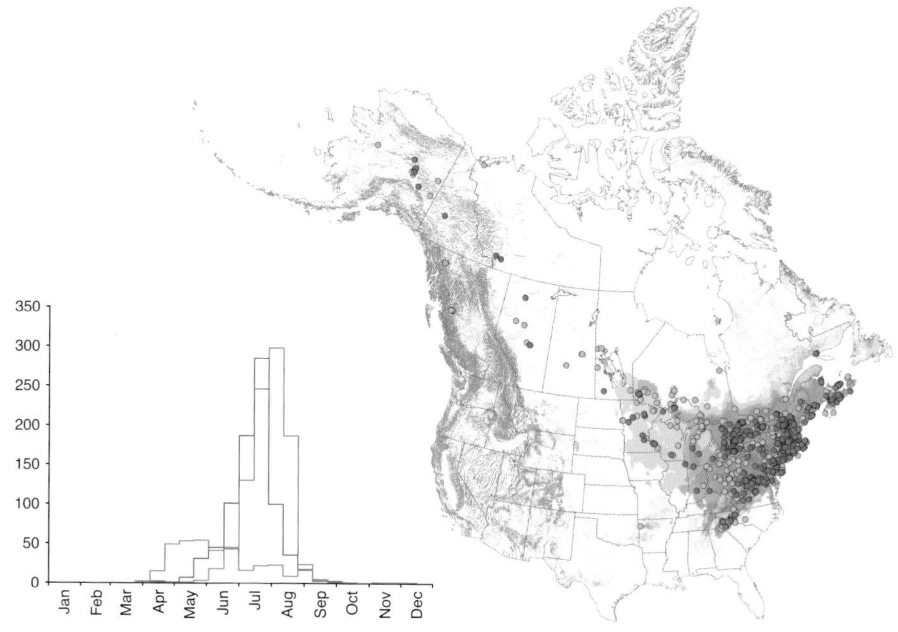
350
300
250
200
150
100
50
0
Jan
Feb
Mar
Apr
May
Jun
Jul
Aug
Sep
Oct
Nov
Dec

HAND CHARACTERS Body size medium (larger than *B. vagans*): queen 18–20 mm (0.70–0.80 inch), worker 11–14 mm (0.44–0.56 inch). Hair long. Head length medium with the *cheek* (oculo-malar area) *just longer than broad* (contrast *B. sandersoni, B. griseocollis, B. rufocinctus, B. affinis*), midleg basitarsus with the back far corner rounded, hindleg tibia outer surface flat without long hair but with long fringes at the sides, forming a pollen basket (corbicula). Hair of the face black or with yellow hairs intermixed, upperside of the thorax between the wings yellow without an obvious spot of black hair or a black band and with only a very few inconspicuous black hairs (contrast most *B. sandersoni*), but the *sides of the thorax often extensively black* (contrast *B. impatiens, B. bimaculatus, B. vagans, B. sandersoni*) although sometimes entirely yellow, metasomal T3 often with some yellow hair intermixed in the middle (contrast *B. vagans, B. sandersoni*), T5–6 often (but not always) with some white hair at least along the back margin (T5) and at the sides (T6). Pale hair of the head, thorax, and T1–3 sometimes brownish rather than yellow (contrast *B. impatiens, B. bimaculatus, B. vagans, B. sandersoni*). Male 13–14 mm (0.51–0.55 inch). Eye similar in size and shape to the eye of any female bumble bee (contrast *B. griseocollis, B. rufocinctus*). Antenna of medium length, flagellum 3× longer than the scape (contrast *B. affinis*). Hair color pattern similar to the queen/worker, but varying to almost completely yellow, often with the sides of the thorax and metasomal T1–3 yellow, T4–6 predominantly black.
MICROSCOPIC CHARACTERS Queen metasomal *T6 with a broadly rounded lump or boss just before the tip* (contrast *B. impatiens, B. bimaculatus, B. vagans, B. sandersoni*). Male genitalia with the penis valve sickle-shaped, the back-curved "sickle" short, flat, and broadly triangular with the sides converging consistently to a rounded point, gonostylus inner (medial) edge nearly straight but *at its base with a very deep notch 1× as broad and half as deep as the breadth of the adjacent penis-valve head, separating the inner (medial) edge of the gonostylus from the inner edge of the gonocoxa* (contrast *B. impatiens, B. bimaculatus, B. vagans, B. sandersoni* without this notch).

OCCURRENCE

RANGE AND STATUS Canadian Maritimes and eastern US in Eastern Temperate Forest and Boreal Forest regions, south in a narrow band at higher elevations along the Appalachian Mountains, west through the Canadian Great Plains, and uncommon in the Tundra/Taiga of Canada and AK.
HABITAT Close to or within wooded areas, urban parks and gardens, wetlands.
EXAMPLE FOOD PLANTS *Campanula, Cirsium, Hypericum, Lonicera, Melilotus, Penstemon, Pontederia, Tilia.*
BEHAVIOR Nests underground. Males patrol circuits in search of mates. One of the species with colonies that end earliest in the summer.
PARASITISM BY OTHER BEES It is likely that this species is a host to *B. flavidus (=fernaldae).*

BOMBUS VAGANS SMITH, 1854
HALF-BLACK BUMBLE BEE
(INCLUDING *BOLSTERI*, ?*COCKERELLI*)

BELOW: **Male *Bombus vagans*.** SCO
INSET: **Worker *Bombus vagans*.** LR

IDENTIFICATION

Eastern and northern, medium long-tongued species. Most similar to *B. impatiens*, *B. bimaculatus*, *B. sandersoni*, and *B. perplexus* (see also *B. rufocinctus*, *B. griseocollis*, and *B. affinis*). The female color patterns with many black hairs intermixed throughout upperside of the thorax and yellow hair on metasomal T5 are found in NM (the color pattern named *cockerelli*), and even darker female color patterns are found in NL (named *bolsteri*). These color patterns also have yellow hairs laterally on T3–4. Evidence from DNA barcodes supports a close relationship between *B. vagans* and *bolsteri*, with only minor differences, as what can be considered parts of a single species. So far it has not been possible to obtain DNA barcodes of *cockerelli* from which to assess whether it is conspecific, the most likely interpretation from female morphology.

HAND CHARACTERS Body size medium (larger than *B. sandersoni*): queen 17–21 mm (0.65–0.83 inch), worker 11–14 mm (0.43–0.54 inch). Hair long. Head length medium, with the

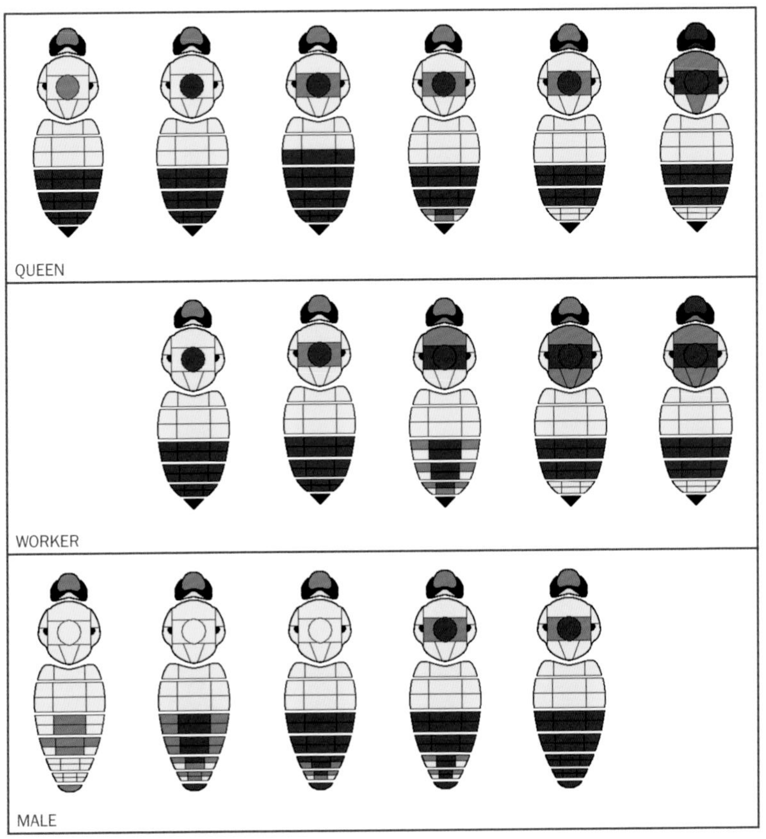
QUEEN
WORKER
MALE

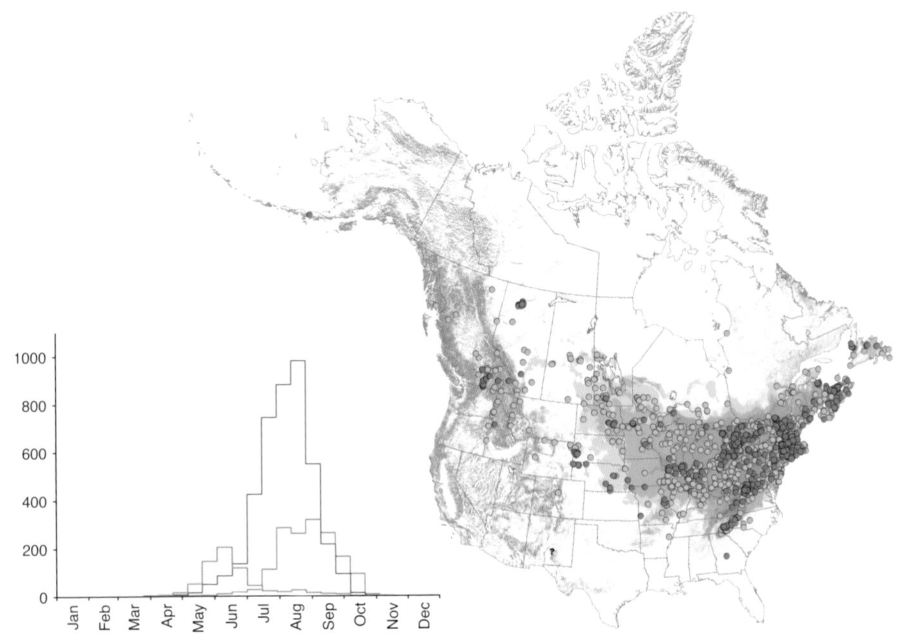
1000
800
600
400
200
0
Jan
Feb
Mar
Apr
May
Jun
Jul
Aug
Sep
Oct
Nov
Dec

cheek (oculo-malar area) *just longer than broad* (contrast *B. sandersoni, B. griseocollis, B. rufocinctus, B. affinis*), midleg basitarsus with the back far corner rounded, hindleg tibia outer surface flat without long hair but with long fringes at the sides, forming a pollen basket (corbicula). Hair of the face black, usually with inconspicuous shorter yellow hairs intermixed around the base of the antenna, upperside of the head with yellow hairs predominant centrally even if sometimes with many black hairs intermixed, upperside of the thorax yellow with a black spot between the wings, although the spot may be small and intermixed with yellow in the queen but larger and more diffuse in the worker, sometimes forming a band between the wings, *sides of the thorax entirely yellow* (contrast most *B. perplexus*), metasomal *T2 mostly yellow across the entire breadth at the front, sometimes with black along the back margin, sometimes intermixed at the back and at the sides but not at the front* (contrast *B. sandersoni*) unless the yellow band is very narrow (contrast *B. impatiens,* most *B. bimaculatus*), T3–4 at the sides and S3–5 usually black with few or no yellow hairs (contrast *B. flavifrons*) except the T3–4 lateral yellow hairs most numerous when there are extensive black hairs on the upperside of the thorax. Male 11–14 mm (0.42–0.55 inch). Eye similar in size and shape to the eye of any female bumble bee (contrast *B. griseocollis, B. rufocinctus*). Antenna of medium length, flagellum 3× longer than the scape (contrast *B. affinis*). Hair of the face yellow with black intermixed at least around the antennal base, thoracic dorsum usually predominantly yellow or sometimes with black hairs intermixed between the wing bases without a distinct black band between the wings (contrast most *B. sandersoni*) (or a strong black band is likely in NL, NM), metasomal T3 usually black at least medially and posteriorly (contrast most *B. perplexus*), T5–6 often with yellow laterally. Rarely T5–6 with orange-red hair, although this may be from damage at the pupal stage.
MICROSCOPIC CHARACTERS Queen metasomal T6 flat or with a short, weak lengthwise ridge just before the tip (contrast *B. perplexus*). Male *antennal segments A3–4 with the back edges without fringing patches of short hair* (contrast *B. sandersoni*), antennal segment A3 long, length nearly 2× its maximum breadth, almost as long as antennal segment A5. Genitalia with the penis valve sickle-shaped, the back-curved "sickle" short and broad, less than 2× longer than the breadth of the broadest part and similar in breadth to the adjacent neck of the penis-valve head, the "sickle" flattened and broadly rounded at the tip (contrast *B. sandersoni*), the penis-valve angle on the underside of the shaft and to the side closer to the base than to the penis-valve head, gonostylus inner (medial) edge nearly straight, the margin thin with a short indistinct parallel submarginal groove, volsella narrow at the far end.

OCCURRENCE

RANGE AND STATUS NL and eastern US in Eastern Temperate Forest and Boreal Forest regions, south in a narrow band at higher elevations along the Appalachian Mountains, west through the northern Great Plains, and in the Mountain West of BC, WA, with the taxon *cockerelli* in NM.
HABITAT A wide variety of habitats, including forests, wetlands, urban parks and gardens.
EXAMPLE FOOD PLANTS *"Aster", Chelone, Cirsium, Impatiens, Lonicera, Monarda, Prunella, Solidago, Trifolium, Vicia, Viola.*
BEHAVIOR Nests mostly underground, although occasionally aboveground. Males patrol circuits in search of mates.
PARASITISM BY OTHER BEES Host to *B. citrinus*, confirmed breeding record.

BOMBUS SANDERSONI FRANKLIN, 1913
SANDERSON BUMBLE BEE

Female *Bombus sandersoni*. MM

Male *Bombus sandersoni*. MM

IDENTIFICATION

Eastern and northern, medium-tongued species. Most similar to *B. impatiens, B. bimaculatus, B. perplexus, B. vagans,* and *B. frigidus* (see also *B. griseocollis, B. rufocinctus,* and *B. affinis*). This species can be difficult to identify and has until recently often been mistaken for other species. Evidence from DNA barcodes supports a close relationship with *B. mixtus.*

HAND CHARACTERS Body size small (smaller than *B. vagans*): queen 15–17 mm (0.58–0.65 inch), worker 10–14 mm (0.40–0.53 inch). Hair short and even. Head length medium with the *cheek* (oculo-malar area) *as long as broad or just shorter than broad* (contrast *B. vagans, B. griseocollis, B. rufocinctus, B. affinis*), midleg basitarsus with the back far corner rounded, hindleg tibia outer surface flat without long hair but with long fringes at the sides, forming a pollen basket (corbicula). Hair of the face and the upperside of the head black or often with yellow hairs, sides of the thorax entirely yellow (contrast *B. perplexus*), upperside of the thorax often with a black band between the wings (contrast *B. perplexus,* most *B. vagans*), corbicular fringes black (contrast *B. frigidus*) or with brownish-orange tips, metasomal *T2 often mostly yellow* (contrast *B. impatiens,* most *B. bimaculatus*) *at the sides and front, often with a few scattered black hairs* (contrast *B. vagans*), T3 black across the entire breadth at least at the front (contrast most *B. perplexus*), T3–4 at the sides and S3–5 black with few or no yellow hairs (contrast *B. flavifrons*), T5 black, orange-brown, yellow, or rarely white, with a few black hairs intermixed. Male 12–15 mm (0.48–0.57 inch). Eye similar in size and shape to the eye of any female bumble bee (contrast *B. griseocollis, B. rufocinctus*). Antenna of medium length, flagellum 3× longer than the scape (contrast *B. affinis*). Hair color pattern similar to the queen/worker.

MICROSCOPIC CHARACTERS Queen metasomal T6 flat or with a short, weak lengthwise ridge just before the tip (contrast *B. perplexus*). Male *antennal segments A3–4 with the back edges with patches of dense short fringing hair but segments A4–12 without similar obvious patches* (contrast *B. mixtus, B. frigidus, B. vagans, B. perplexus*), antennal segment A3 short, length less than 1.5× its maximum breadth, much shorter than antennal segment A5. Hind basitarsus outer surface at the near end and away from the back edge with hairs as long as half the breadth of the basitarsus (contrast *B. mixtus*). Genitalia with the penis valve sickle-shaped, the back-curved "sickle" short

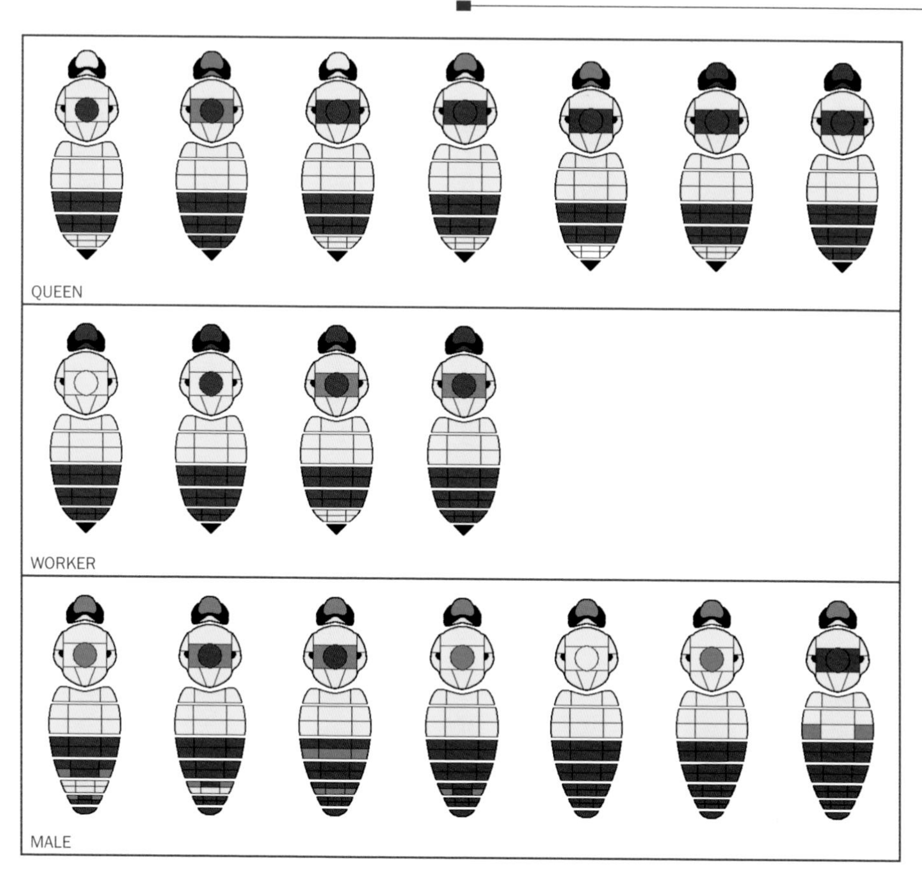
QUEEN
WORKER
MALE

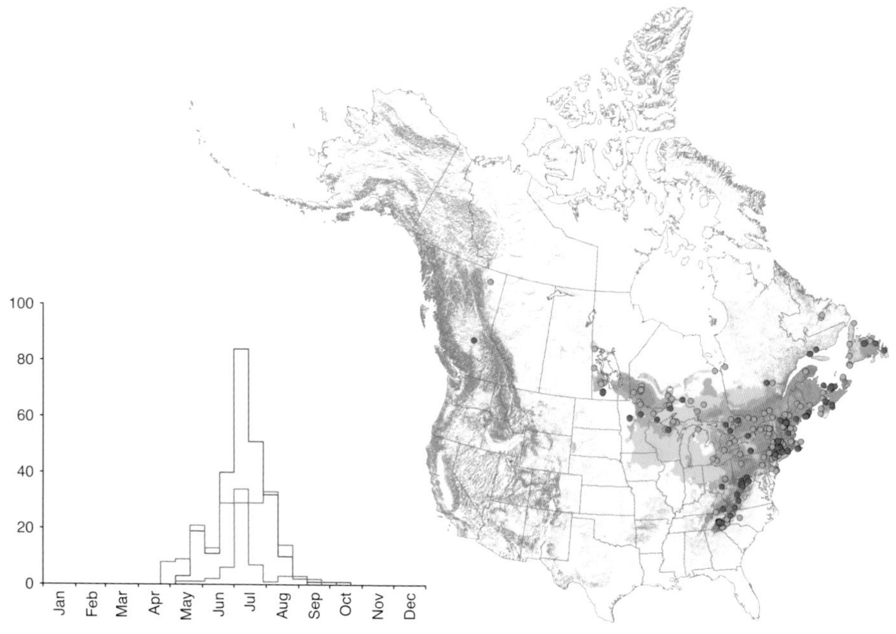
100
80
60
40
20
0
Jan
Feb
Mar
Apr
May
Jun
Jul
Aug
Sep
Oct
Nov
Dec

and broad, broader than the adjacent neck of the penis-valve head (contrast *B. vagans*, *B. frigidus*), the "sickle" flattened and rounded at the tip, the penis-valve angle on the underside of the shaft and to the side located at the midpoint of the penis-valve shaft length (contrast with closer to the base in *B. vagans*, *B. frigidus*), gonostylus inner (medial) edge nearly straight, the margin thick with a distinct long parallel sub-marginal groove.

Female *Bombus sandersoni*. CS

OCCURRENCE

RANGE AND STATUS Canadian Maritimes and eastern US in Eastern Temperate Forest and Boreal Forest regions, south in a narrow band at higher elevations along the Appalachian Mountains, west through the Canadian Great Plains, and uncommonly in the Tundra/Taiga of Canada.

HABITAT In or near wooded areas.

EXAMPLE FOOD PLANTS *Cimicifuga, "Epilobium", Kalmia, Lonicera, Lyonia, Melilotus, Monarda, Rubus, Vaccinium.*

BEHAVIOR Nests underground. Males patrol circuits in search of mates.

PARASITISM BY OTHER BEES Unknown.

BOMBUS JONELLUS (KIRBY, 1802)
WHITE TAIL BUMBLE BEE (INCLUDING *ALBOANALIS*)

Female *Bombus jonellus*. GH

Male (Britain) *Bombus jonellus*. SF

IDENTIFICATION

Northwestern, medium-tongued species. Most similar to some *B. vagans, B. sandersoni, B. perplexus, B. frigidus,* and *B. mixtus* (see also *B. cryptarum)*. Evidence from DNA barcodes supports a close relationship between the bees named *alboanalis* from North America and *B. jonellus* from northern Europe and Asia, which appear to be parts of the same species.

HAND CHARACTERS Body size small: queen 16 mm (0.62–0.64 inch), worker 10–11 mm (0.38–0.42 inch). Hair long. Head length medium, with the cheek (oculo-malar area) as long as broad (contrast *B. cryptarum*), midleg basitarsus with the back far corner rounded, hindleg tibia outer surface flat without long hair but with long fringes at the sides, forming a pollen basket (corbicula). Hair *of the face black with a tuft of yellow hair intermixed at the base of the antenna* (contrast *B. mixtus*), center upperside of the head yellow, upperside of the thorax with the front pale band without black hair intermixed, a distinct black band between the wings (contrast *B. perplexus*), corbicular fringes black, orange, or black with pale tips, metasomal T2 without scattered black hairs at the sides near the front except sometimes in the extreme corners, *T4–5 usually both white or slightly yellowish* (contrast *B. vagans, B. sandersoni, B. frigidus),* or sometimes T4 black in the front half. Male 10–12 mm (0.40–0.46 inch). Eye similar in size and shape to the eye of any female bumble bee. Antenna of medium length, flagellum 2.5–3× longer than the scape (contrast *B. cryptarum*). Hair color pattern similar to the queen/worker, with metasomal T5–6 white.

MICROSCOPIC CHARACTERS Male *antennal segments A3–4 with the back edges without dense fringing patches of short hair* (contrast *B. sandersoni*), at most with a very few short bristles, antennal segment A3 short, length less than 1.5× its maximum breadth, much shorter than antennal segment A5. Genitalia with penis valve sickle-shaped, the back-curved "sickle" short and broad, less than 2× longer than the breadth of the broadest part and similar in breadth to the adjacent neck of the penis-valve head, the "sickle" flattened and truncated at the tip, the penis-valve angle on the underside of the shaft and to the side closer to the base than to the penis-valve head, gonostylus inner (medial) edge nearly straight, the margin thick with a distinct long parallel submarginal groove.

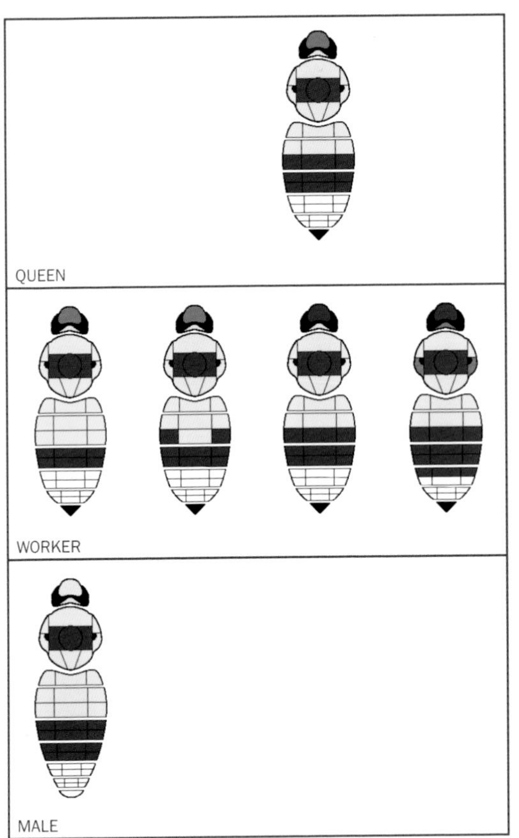

OCCURRENCE

RANGE AND STATUS Tundra/Taiga region from Hudson Bay, MB, west to AK. Also in Europe and Asia.

HABITAT Tundra/Taiga.

EXAMPLE FOOD PLANTS *Arctostaphylos, "Epilobium", Hedysarum, Trifolium, Vaccinium, Vicia.*

BEHAVIOR Nests underground. Males patrol circuits in search of mates.

PARASITISM BY OTHER BEES Unknown.

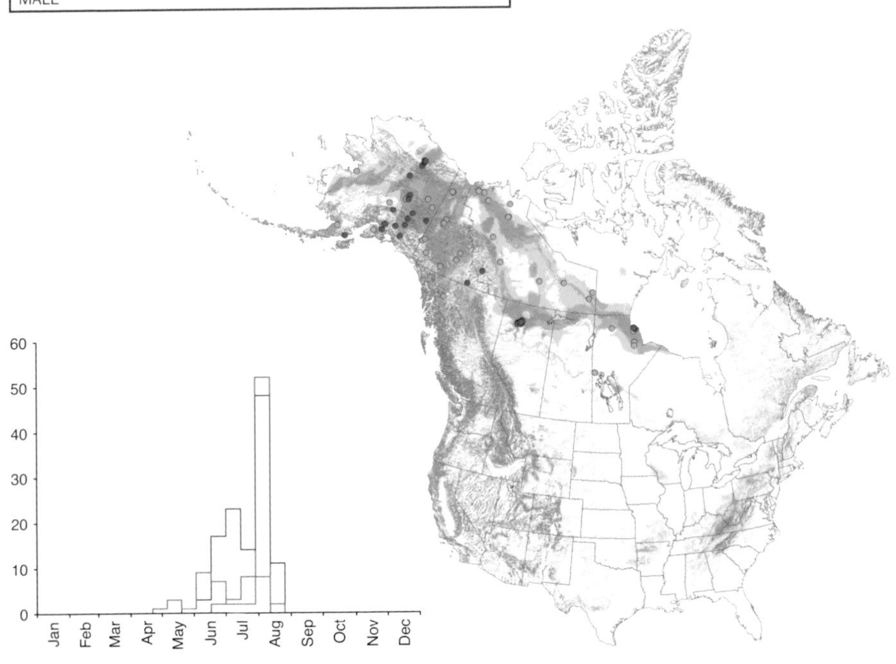

BOMBUS FRIGIDUS SMITH, 1854
FRIGID BUMBLE BEE

Male *Bombus frigidus*. RT

Worker *Bombus frigidus*. JF

IDENTIFICATION

Northern and western mountain, medium-tongued species. Most similar to *B. polaris, B. balteatus, B. mixtus, B. sandersoni, B. jonellus,* and *B. vagans.*

HAND CHARACTERS Body size medium: queen 17–19 mm (0.65–0.74 inch), worker 8–11 mm (0.32–0.45 inch). Hair long. Head length medium with the cheek (oculo-malar area) as long as broad (contrast *B. balteatus*), midleg basitarsus with the back far corner rounded, hindleg tibia outer surface flat without long hair but with long fringes at the sides, forming a pollen basket (corbicula). Hair of the face black or with some short yellow hairs, upperside of the thorax with the *front pale band usually without black hair extensively intermixed* (contrast *B. mixtus*), sides of the thorax yellow or rarely extensively black, corbicular fringes usually extensively pale orange (contrast *B. sandersoni*) but occasionally entirely black, metasomal T2 usually without scattered black hairs at the sides near the front, *T3 without yellow* (contrast *B. balteatus*, many *B. polaris*), *T5 orange* (contrast *B. sandersoni, B. jonellus, B. vagans*) although sometimes this is very pale. Male 10–15 mm (0.40–0.57 inch). Eye similar in size and shape to the eye of any female bumble bee. Antenna of medium length, flagellum 3× longer than the scape. Hair color pattern similar to the queen/worker.

MICROSCOPIC CHARACTERS Queen/worker mandible with only a shallow notch in front of the back tooth (contrast *B. polaris, B. balteatus*), clypeus from its central area toward the labrum with a few very large punctures with large intervening bare smooth areas with only a few obscure micropunctures (contrast *B. polaris*). Male antennal segments A3–4 with the back edges without dense fringing patches of short hair (contrast *B. sandersoni, B. mixtus*), at most with a very few short hairs, antennal segment A3 short, length less than 1.5× its maximum breadth, much

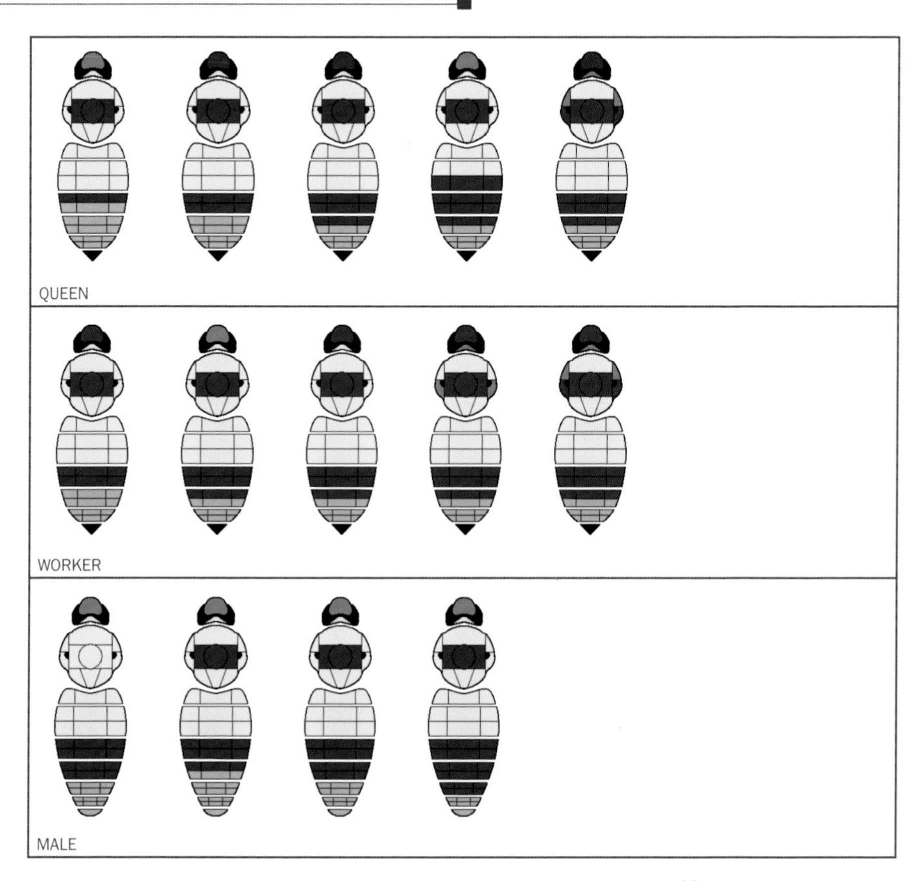
QUEEN
WORKER
MALE

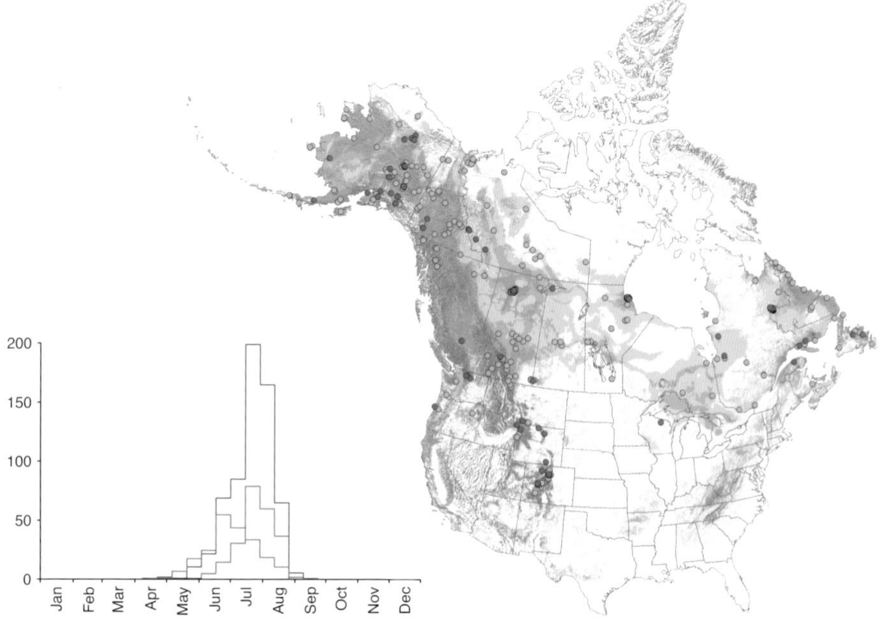
200
150
100
50
0
Jan
Feb
Mar
Apr
May
Jun
Jul
Aug
Sep
Oct
Nov
Dec

shorter than antennal segment A5, hind basitarsus outer surface at the near end and away from the back edge with no hairs quite as long as half the breadth of the basitarsus (contrast *B. sandersoni*). Genitalia with the penis valve sickle-shaped (contrast *B. polaris, B. balteatus*), the back-curved "sickle" short and broad, less than 2× longer than the breadth of the broadest part and similar in breadth to the adjacent neck of the penis-valve head, the "sickle" flattened and broadest at the midpoint of its length (contrast *B. sandersoni, B. mixtus*), the penis-valve angle on the underside of the shaft and to the side closer to the base than to the penis-valve head (contrast *B. sandersoni, B. mixtus*), gonostylus with the inner (medial) edge nearly straight, the margin thick with a distinct long parallel submarginal groove, volsella broad at the far end (contrast *B. vagans*).

OCCURRENCE

RANGE AND STATUS Tundra/Taiga and Boreal Forest regions from eastern maritime Canada west to AK, and in the Mountain West south to OR, CO.

HABITAT Tundra/Taiga and mountain meadow.

EXAMPLE FOOD PLANTS *"Epilobium", Hedysarum, Lupinus, Potentilla, Salix, Trifolium, Vaccinium.*

BEHAVIOR Nests often aboveground, sometimes underground. Males patrol circuits in search of mates.

PARASITISM BY OTHER BEES Unknown.

BOMBUS MIXTUS CRESSON, 1878
FUZZY-HORNED BUMBLE BEE

Female *Bombus mixtus*. RT

Male *Bombus mixtus*. BK

IDENTIFICATION

Northern and western mountain, medium-tongued species. Most similar to *B. frigidus, B. sandersoni, B. polaris, B. balteatus,* and *B. melanopygus. B. mixtus* may be a junior synonym of Kirby's *B. praticola*, although the type specimen for that name is lost. Despite the precedence of

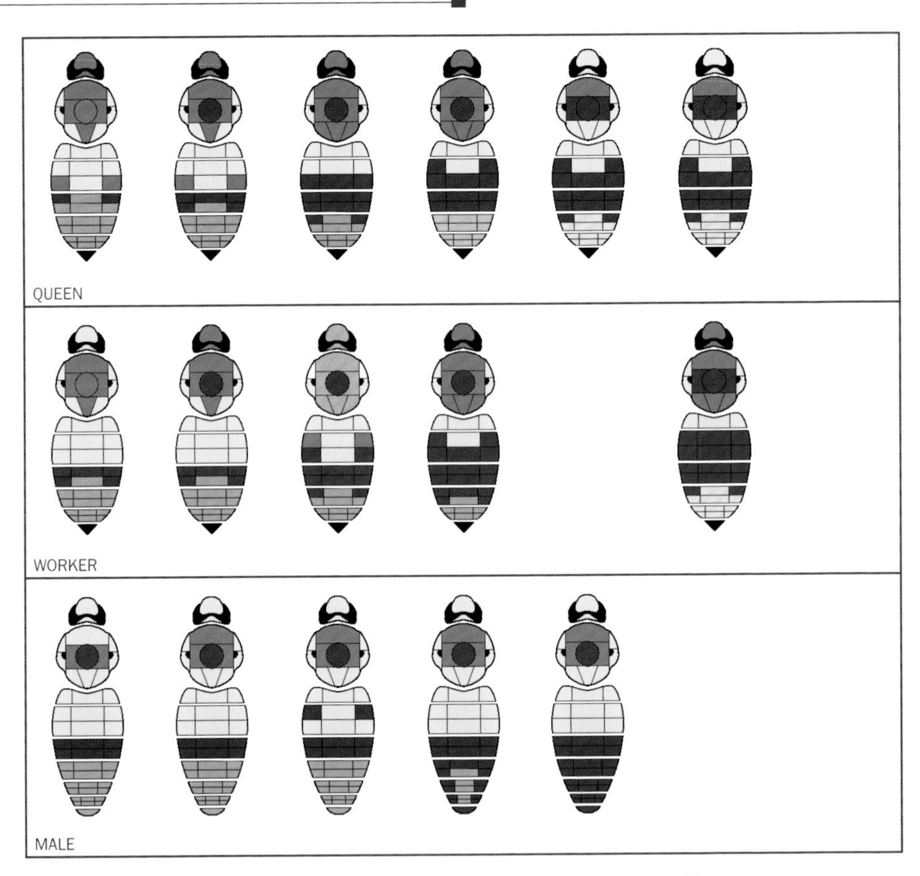
QUEEN
WORKER
MALE

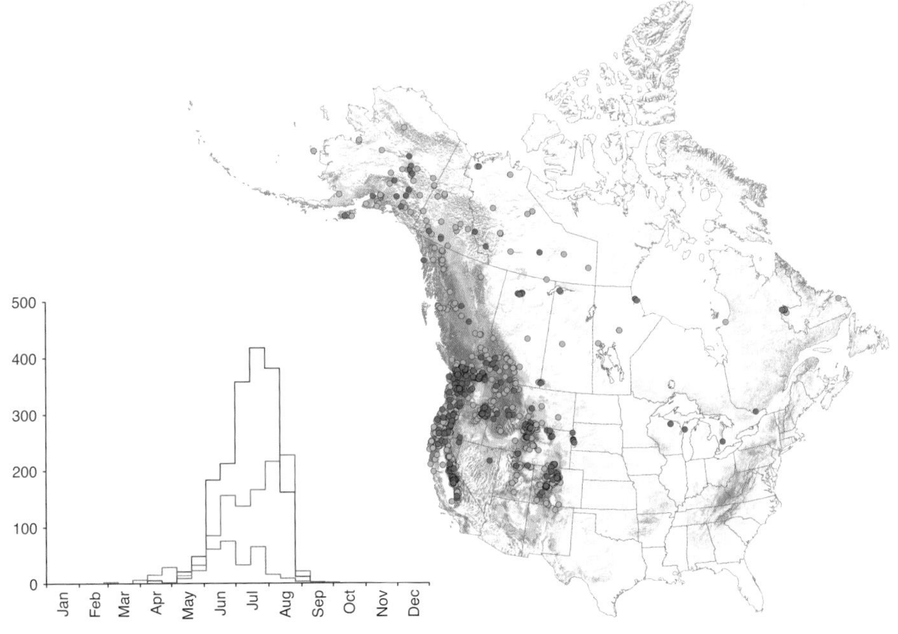
500
400
300
200
100
0
Jan
Feb
Mar
Apr
May
Jun
Jul
Aug
Sep
Oct
Nov
Dec

the earlier name (if the type is correctly identified), the name *B. mixtus* has been in common use for this species since 1950 and we know of only one publication using the name *B. praticola*. We suggest that, for stability, we should continue to use the name *B. mixtus*. Tail color paler and the metasoma with more extensive black at the southern end of the range.

HAND CHARACTERS Body size small: queen 15–17 mm (0.58–0.68 inch), worker 10–14 mm (0.37–0.56 inch). Hair long and uneven. Head length medium with the cheek (oculo-malar area) as long as broad (contrast *B. balteatus*), midleg basitarsus with the back far corner rounded, hindleg tibia outer surface flat without long hair but with long fringes at the sides, forming a pollen basket (corbicula). Hair *on the face* (contrast *B. jonellus*), *upperside of the head, and upperside of the thorax with the front pale band yellow (or sometimes grayish) usually with many black hairs intermixed, any front edge of a black band between the wings not sharply defined* (contrast *B. frigidus, B. polaris, B. balteatus*), corbicular fringes usually predominantly pale orange (contrast *B. sandersoni*) or at least black with orange tips, tail usually with very pale orange hair, including metasomal T5 orange (contrast *B. bifarius, B. flavifrons*) although occasionally yellowish. Male 11–14 mm (0.43–0.53 inch). Eye similar in size and shape to the eye of any female bumble bee. Antenna moderately long and thick, flagellum 3.3× longer than the scape. Hair color pattern similar to the queen/worker, or T3–7 rarely all black.

MICROSCOPIC CHARACTERS Queen/worker mandible with only a shallow notch in front of the back tooth (contrast *B. polaris, B. balteatus*). Male antennal segments A3–13 unusually broad, *antennal segments A3–11 with the back edges with fringing patches of short to medium length hair* (contrast *B. frigidus, B. sandersoni, B. polaris, B. balteatus*), antennal segment A3 short, length less than 1.5× its maximum breadth, much shorter than antennal segment A5. Genitalia with penis valve sickle-shaped (contrast *B. polaris, B. balteatus*), the back-curved "sickle" moderately broad, less than 2× longer than breadth of the broadest part and similar in breadth to the adjacent neck of the penis-valve head, the "sickle" flattened and truncate and broadest near the tip, the penis-valve angle on the underside of the shaft and to the side at the midpoint of the length from the base to the penis-valve head (contrast *B. sandersoni*), gonostylus inner (medial) edge nearly straight, the margin thick with a distinct long parallel submarginal groove, volsella broad at the far end (contrast *B. vagans*).

OCCURRENCE

RANGE AND STATUS Mountain West from CA in the southern Sierra Nevada to BC, in the Rocky Mountains to CO and NM, and in the Tundra/Taiga region of AK and western Canada, with disjunct populations in the upper US Midwest, northern ON, and Maritime NB. From sea level to about 2,100 m. Moderately common throughout its range.

HABITAT Open grassy areas, chaparral and shrub areas, mountain meadows.

EXAMPLE FOOD PLANTS *Ceanothus, "Epilobium", Monardella, Penstemon, Phacelia, Rhododendron, Senecio.*

BEHAVIOR Nest locations variable, many on the surface of the ground, but some above- or belowground. Males patrol circuits in search of mates.

PARASITISM BY OTHER BEES Unknown.

BOMBUS TERNARIUS SAY, 1837
TRI-COLORED BUMBLE BEE

Female *Bombus ternarius*. LR

Male *Bombus ternarius*. LR

IDENTIFICATION

Eastern and northern, medium-tongued species. Most similar to *B. huntii, B. sylvicola, B. melanopygus,* and *B. bifarius* (see also *B. rufocinctus*). Evidence from DNA barcodes supports subgroups previously identified as parts of this species as unusually divergent from one another in the COI gene. This species replaces *B. huntii* in the north.

HAND CHARACTERS Body size medium: queen 17–19 mm (0.67–0.74 inch), worker 9–13 mm (0.36–0.51 inch). Hair short and even (contrast *B. sylvicola*). Head length medium with the cheek (oculo-malar area) very slightly shorter than broad (contrast *B. rufocinctus*), midleg basitarsus with the back far corner rounded, hindleg tibia outer surface flat without long hair but with long fringes at the sides, forming a pollen basket (corbicula). Hair of the face and upperside of the head often predominantly black with patches of yellow, or sometimes with yellow more intermixed especially for the worker, upperside of the thorax with the yellow front band sometimes with sparse black hair intermixed especially for the worker but the front edge of the black band between the wings always sharply defined, the black band between the wings *extending backward in a V shape along the middle (of the scutellum) for at least a short distance and often as far as the back edge* (contrast *B. huntii, B. sylvicola, B. melanopygus*), corbicular fringes black (contrast *B. bifarius*), metasomal T1 predominantly yellow, T2 orange at most with a few black hairs at the front in the middle, T3 orange, T4 yellow, T5 black. Male 10–14 mm (0.38–0.54 inch). Eye similar in size and shape to the eye of any female bumble bee (contrast *B. rufocinctus*). Antenna of medium length, flagellum 2.5–3× longer than the scape. Hair color pattern similar to the queen/worker, but the black V on the upperside of the thorax at the back (scutellum) less obvious than in the queen/worker.

MICROSCOPIC CHARACTERS Male genitalia with the penis valve sickle-shaped, the back-curved "sickle" long and very narrow, at least 3× longer than the breadth of the further half and less than half the breadth of the adjacent neck of the penis-valve head, the "sickle" scarcely flattened and about 2× broader than thick, almost spinelike, *the penis-valve angle on the underside of the shaft and to the side located more than 0.33× the distance from the base to the penis-valve head* (contrast

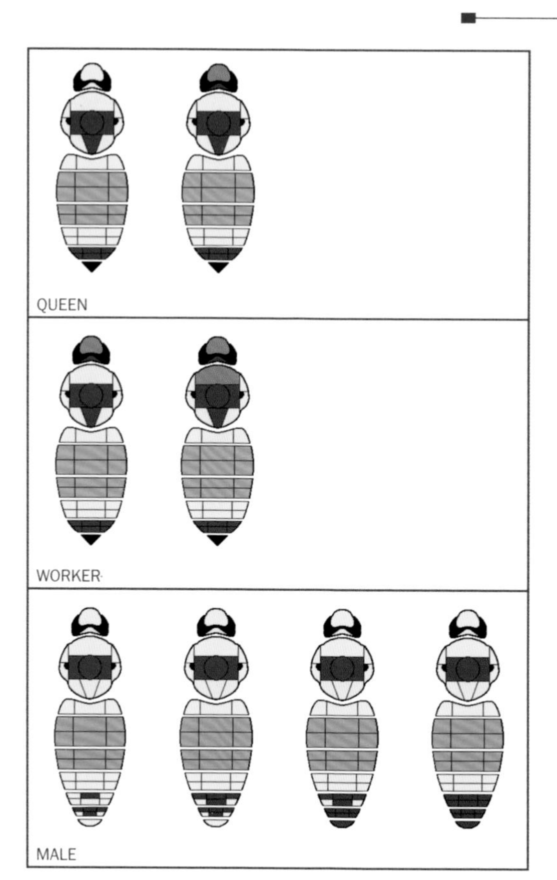

B. sylvicola), gonostylus as long as broad (contrast *B. sylvicola*), the inner (medial) edge strongly concave.

OCCURRENCE

RANGE AND STATUS Canadian Maritimes and US Eastern Temperate Forest and Boreal Forest regions, west through ND and Canadian Great Plains, scattered in the US Mountain West to BC. Common in the northeast Boreal Forest.

HABITAT Close to or within woodland, wetlands.

EXAMPLE FOOD PLANTS *Asclepias, Claytonia, Eupatorium, Rhododendron, Rubus, Solidago, Symphyotrichum, Taraxacum, Trifolium, Vaccinium.*

BEHAVIOR Nests often underground. Males patrol circuits in search of mates.

PARASITISM BY OTHER BEES Host to *B. insularis*, confirmed breeding record.

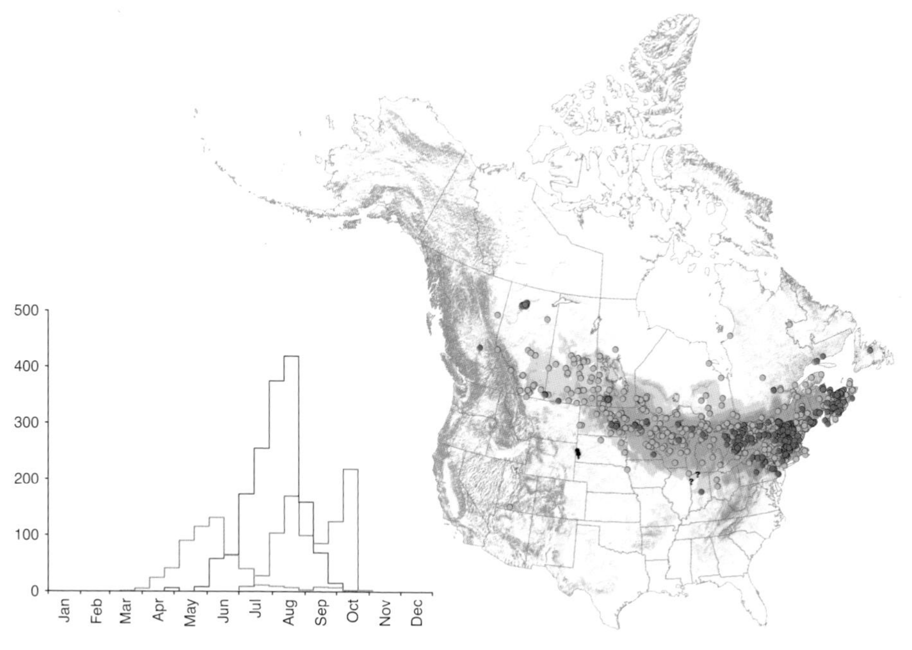

BOMBUS HUNTII GREENE, 1860
HUNT BUMBLE BEE

LEFT: Queen *Bombus huntii.* MH
ABOVE: Queen and Male *Bombus huntii.* NS

IDENTIFICATION

Western interior, medium-tongued species. Most similar to *B. ternarius, B. sylvicola, B. melanopygus,* and *B. bifarius* (see also *B. rufocinctus*). This species replaces *B. ternarius* in the interior of the western US.

HAND CHARACTERS Body size medium: queen 19–20 mm (0.73–0.79 inch), worker 11–14 mm (0.42–0.54 inch). Hair short and even (contrast *B. sylvicola*). Head length medium with the cheek (oculo-malar area) as long as broad (contrast *B. rufocinctus*), midleg basitarsus with the back far corner rounded, hindleg tibia outer surface flat without long hair but with long fringes at the sides, forming a pollen basket (corbicula). Hair of the face and the upperside of the head predominantly yellow with few black hairs (contrast *B. ternarius*), upperside of the thorax with the front yellow band without black hairs intermixed (contrast *B. melanopygus*, some *B. bifarius*), sides of the thorax mostly yellow with black only below and at the back, *upperside of the thorax at the back (scutellum) entirely yellow* (contrast *B. ternarius*), corbicular fringes black (contrast *B. bifarius*). Male 9–13 mm (0.37–0.52 inch). Eye similar in size and shape to the eye of any female bumble bee (contrast *B. rufocinctus*). Antenna of medium length, flagellum 2.5–3× longer than the scape. Hair color pattern similar to the queen/worker, but the upperside of the thorax often with many yellow hairs intermixed in the black band between the wings (contrast *B. ternarius*).

MICROSCOPIC CHARACTERS Male genitalia with the penis valve sickle-shaped, the back-curved "sickle" long and very narrow, at least 3× longer than broad and less than half the breadth of the adjacent neck of the penis-valve head, the "sickle" scarcely flattened and about 2× broader than thick, almost spinelike and not expanded at the tip (contrast *B. sylvicola*), *the penis-valve angle on the underside of the shaft and to the side located at more than 0.33× the distance from the base to the penis-valve head* (contrast *B. sylvicola*), gonostylus as long as broad (contrast *B. sylvicola*), with the inner (medial) edge strongly concave.

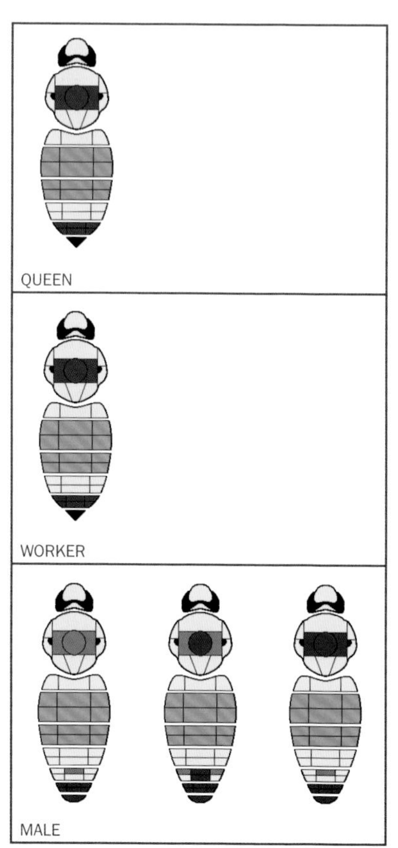

OCCURRENCE

RANGE AND STATUS Mountain West east of the Sierra Nevada and in the Rocky Mountains, adjacent southern Canada, and in the Great Plains of the northern US and southern Canada. Also in Mexico.

HABITAT High desert scrub.

EXAMPLE FOOD PLANTS *Chrysothamnus, Cirsium, Ericameria, Helianthus, Lupinus, Medicago, Melilotus, Penstemon, Phacelia, Ribes, Rudbeckia, Trifolium.*

BEHAVIOR Nests underground. Males patrol circuits in search of mates.

PARASITISM BY OTHER BEES Unknown.

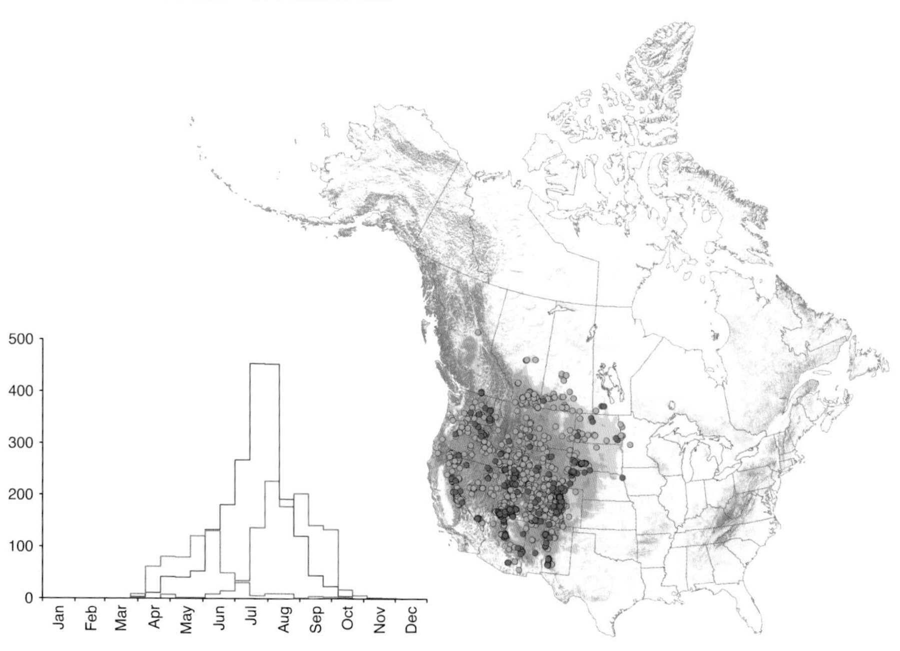

BOMBUS SYLVICOLA KIRBY, 1837

FOREST BUMBLE BEE (INCLUDING *GELIDUS*)

TOP: **Female *Bombus sylvicola*.** RB
ABOVE: **Male *Bombus sylvicola*.** AR

IDENTIFICATION

Northern and western mountain, medium-tongued species. Most similar to *B. melanopygus*, *B. bifarius*, *B. huntii*, *B. ternarius*, *B. sandersoni*, and *B. sitkensis* (see also *B. rufocinctus*). Evidence from DNA barcodes supports a close relationship between *B. sylvicola* and the *B. lapponicus*-complex from the north of Europe and Asia, and *B. sylvicola* may be a part of a broader species. DNA barcode evidence also confirms that the two principal color forms in North America are

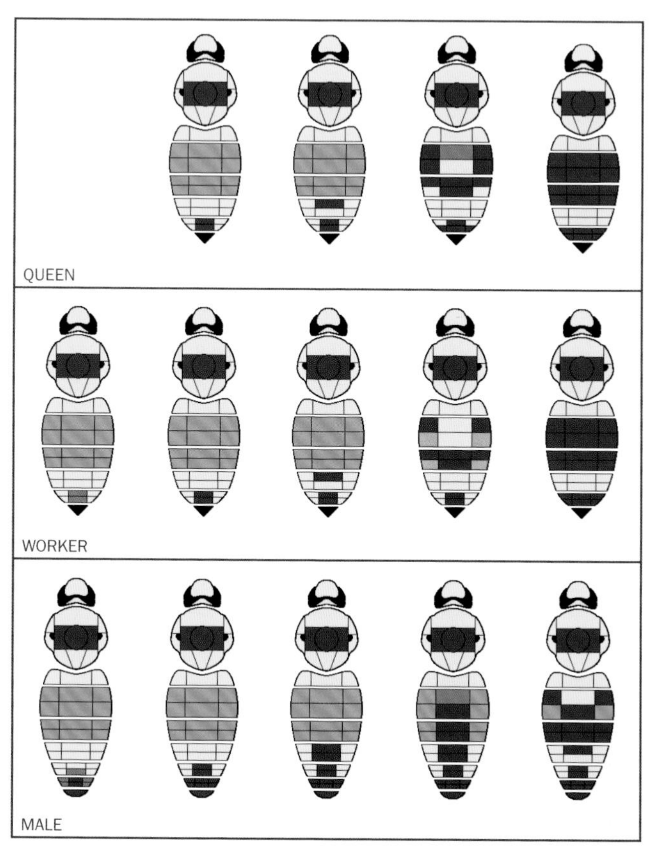
QUEEN
WORKER
MALE

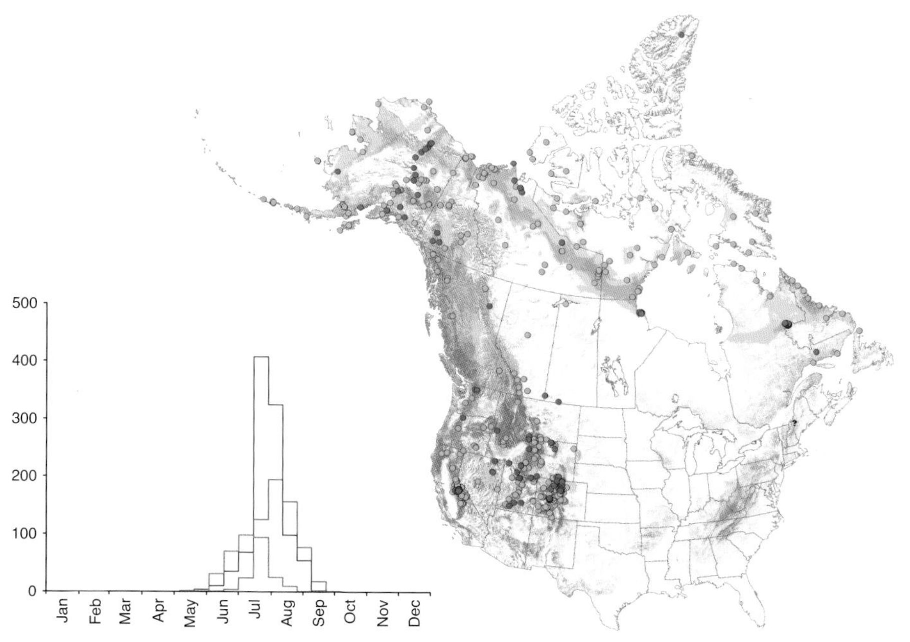
500
400
300
200
100
0
Jan
Feb
Mar
Apr
May
Jun
Jul
Aug
Sep
Oct
Nov
Dec

conspecific: (1) with metasomal T2–3 red (e.g., White Mountains, Great Basin, Rocky Mountains); and (2) with T2–3 predominantly black (Sierra Mountains, CA). In addition, yellow hair is sometimes largely replaced with black on the face and on the sides of the thorax (named *gelidus*, most frequent in AK).

HAND CHARACTERS Body size small: queen 15–17 mm (0.59–0.68 inch), worker 10–14 mm (0.40–0.56 inch). Hair long and uneven (contrast *B. ternarius, B. huntii*). Head length medium with the cheek (oculo-malar area) as long as broad (contrast *B. rufocinctus*), midleg basitarsus with the back far corner rounded, hindleg tibia outer surface flat without long hair but with long fringes at the sides, forming a pollen basket (corbicula). *Hair of the face and the upperside of the head often extensively black with small patches of yellow around the antennal base and above* (contrast *B. bifarius, B. huntii, B. melanopygus*), upperside of the thorax with the front yellow band with or without black hairs, but at most sparsely intermixed, the front edge of the black band between the wings always sharply defined (contrast *B. melanopygus*), *upperside of the thorax at the back (scutellum) with yellow patches at the sides not entirely divided along the midline to the back edge by black* (contrast *B. ternarius, B. bifarius*), metasomal T2 red or sometimes with a narrow to large front middle triangle of black hairs pointing backward and reaching back along the midline to the back edge, T4–5 often predominantly yellow although usually with some black in the middle. Sometimes T2–3 with part or complete replacement of red hair with black and yellow, but if so then T4 is yellow and T5 is extensively black with yellow only at the extreme sides (contrast *B. sandersoni, B. sitkensis*). Male 11–14 mm (0.42–0.54 inch). Eye similar in size and shape to the eye of any female bumble bee (contrast *B. rufocinctus*). Antenna of medium length, flagellum 3× longer than the scape. Hair color pattern similar to the queen/worker, upperside of the thorax sometimes with the black band between the wings with many yellow hairs intermixed.

MICROSCOPIC CHARACTERS Queen/worker feet (tarsi) black like the tibia (contrast *B. bifarius*). Male genitalia with penis valve sickle-shaped, the back-curved "sickle" long and very narrow, *often expanded slightly at the tip* (contrast *B. melanopygus, B. bifarius, B. huntii, B. ternarius, B. sandersoni, B. sitkensis*), the *penis valve with a strong angle on the underside of the shaft and to the side located about 0.25× the distance from base to the penis-valve head* (contrast *B. melanopygus, B. bifarius, B. huntii, B. ternarius, B. sandersoni, B. sitkensis*), gonostylus shorter than broad, the inner (medial) edge concave.

OCCURRENCE

RANGE AND STATUS A boreal-alpine species widespread throughout most of northern North America, from the Mountain West, CO, UT, NV, CA north to AK, and through the Tundra/Taiga region across the continent to NF. In the southern part of its range (CA) it is found mostly above 2,400 m up to 4,200 m. Moderately common.

HABITAT Open grassy areas and mountain meadows.

EXAMPLE FOOD PLANTS *Arenaria, Chrysothamnus, "Epilobium", Haplopappus, Lupinus, Monardella, Petasites, Phyllodoce, Senecio.*

BEHAVIOR Nests mostly underground, or sometimes on the surface. Males patrol circuits in search of mates.

PARASITISM BY OTHER BEES Unknown.

BOMBUS MELANOPYGUS NYLANDER, 1848
BLACK TAIL BUMBLE BEE (INCLUDING *EDWARDSII*)

LEFT: **Female *Bombus melanopygus*.** GZ
ABOVE: **Male *Bombus melanopygus*.** LRE
BELOW: **Male *Bombus melanopygus*.** GZ

IDENTIFICATION

Northern and western, medium-tongued species. Most similar to *B. sylvicola, B. bifarius, B. ternarius, B. huntii, B. mixtus, B. sitkensis,* and *B. sandersoni* (see also *B. rufocinctus*). Evidence from DNA barcodes supports a close relationship between individuals with the two principal color patterns, which are considered conspecific: (1) with metasomal T2–3 red (named *melanopygus*, from most of the range south to CO), and (2) with T2–3 black or black and yellow (named *edwardsii*, from CA and OR).

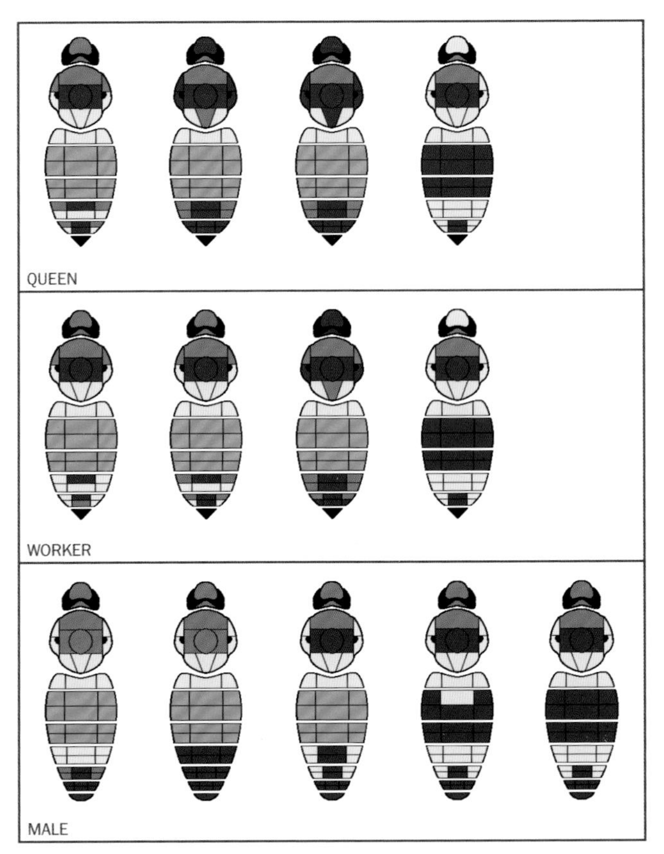
QUEEN
WORKER
MALE

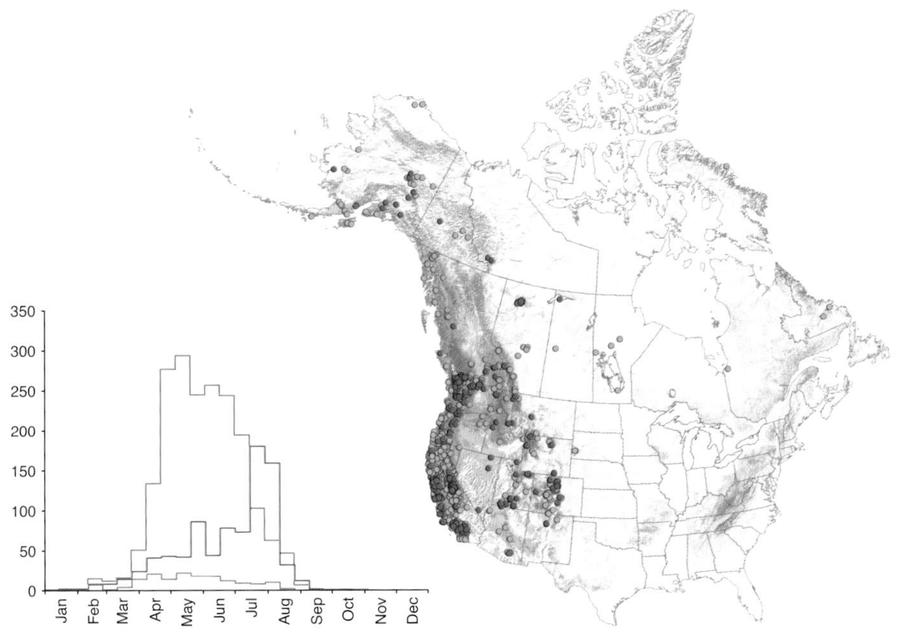
350
300
250
200
150
100
50
0
Jan
Feb
Mar
Apr
May
Jun
Jul
Aug
Sep
Oct
Nov
Dec

HAND CHARACTERS Body size small: queen 16–19 mm (0.63–0.76 inch), worker 10–16 mm (0.41–0.63 inch). Hair short and even (or on Baffin Island longer and uneven). Head length medium with the cheek (oculo-malar area) as long as broad (contrast *B. bifarius, B. rufocinctus*), midleg basitarsus with the back far corner rounded, hindleg tibia outer surface flat without long hair but with long fringes at the sides, forming a pollen basket (corbicula). Hair on the face and upperside of the head yellow, upperside of the thorax with the front pale band with many black hairs densely intermixed (black hairs become more frequent in individuals further north) and on the sides of the thorax in the upper half, the front edge of the black band between the wings often not sharply defined (contrast *B. sylvicola, B. bifarius, B. huntii, B. ternarius*), the pale band at the back of the upperside of the thorax variable from uninterrupted yellow to usually only narrowly and partly divided along the midline by black (except on Baffin Island, contrast *B. bifarius*), hindleg with the corbicular fringes black with at most orange tips, metasomal T2–3 red sometimes with black hairs in the middle and sometimes reaching to the back edge, T4–5 black and often with some yellow (in individuals from the south). If T2–3 are predominantly black, then T4–5 are extensively yellow with T5 broadly yellow at the sides but on the midline narrowly black with a black fringe at the back (contrast *B. bifarius, B. sandersoni, B. sitkensis*). Male 11–14 mm (0.42–0.56 inch). Eye similar in size and shape to the eye of any female bumble bee (contrast *B. rufocinctus*). Antenna of medium length, flagellum 3× longer than the scape. Hair color pattern similar to the queen/worker, upperside of the thorax sometimes with the black band between the wings extensively intermixed with yellow, rarely with metasomal T4–7 black.
MICROSCOPIC CHARACTERS Male genitalia with the penis valve sickle-shaped, the back-curved "sickle" long and narrow, at least 3× longer than broad, but distinctly flattened and more than 3× broader than thick, not expanded at the tip (contrast *B. sylvicola*), the penis-valve angle on the underside of the shaft and to the side only slightly closer to the base than to the penis-valve head (contrast *B. sylvicola*), gonostylus as long as broad (contrast *B. sylvicola*), the inner (medial) edge strongly concave.

OCCURRENCE

RANGE AND STATUS Throughout most of CA except the southeastern deserts, highlands of the Desert West of AZ, NM, UT, NV, scattered over most of the Mountain West, also found sparingly through Tundra/Taiga region of AK and western Canada and the northern Great Plains, with apparently disjunct populations from the Arctic and from northern ON and maritime NB. From sea level to above 2,100 m. Common throughout its range in the west.
HABITAT Open grassy areas, urban parks and gardens, chaparral and shrub areas, mountain meadows.
EXAMPLE FOOD PLANTS *Arctostaphylos, Ceanothus, Ericameria, Eriodyction, Eriogonum, Haplopappus, Lupinus, Penstemon, Rhododendron, Salix, Salvia, Trifolium, Vaccinium, Wyethia.*
BEHAVIOR Nests underground or aboveground in birdhouses, insulation in buildings. Males patrol circuits in search of mates. This species is one of the earliest to start nesting and to produce males.
PARASITISM BY OTHER BEES Unknown.

BOMBUS BIFARIUS CRESSON, 1878
TWO FORM BUMBLE BEE (INCLUDING *NEARCTICUS*)

ABOVE: **Queen and male *Bombus bifarius*.** DK
INSET: **Male *Bombus bifarius*.** VK

IDENTIFICATION

Western, medium-tongued species. Most similar to *B. ternarius, B. huntii, B. sylvicola, B. melanopygus, B. sitkensis,* and *B. sandersoni* (see also *B. rufocinctus*). Evidence from DNA barcodes supports a close relationship between two divergent groups with different color patterns: (1) with metasomal T2–3 extensively red (named *bifarius*, from CO north to BC); and (2) with T2–3 extensively black (named *nearcticus*, predominant in BC, AK, YT, AB, CA, OR, ID, MT, WA).

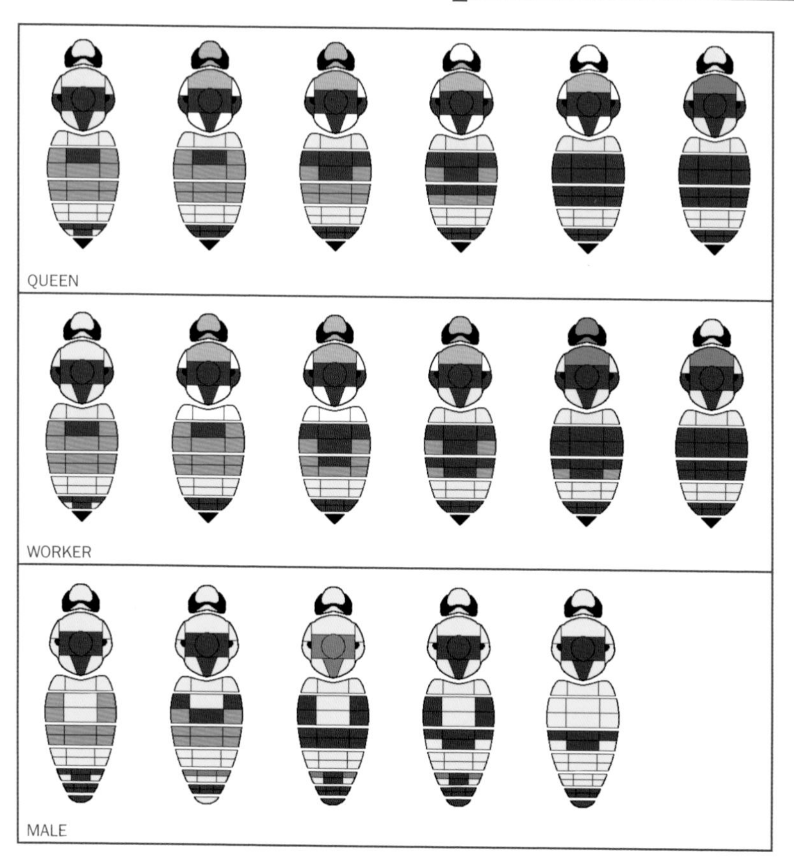
QUEEN
WORKER
MALE

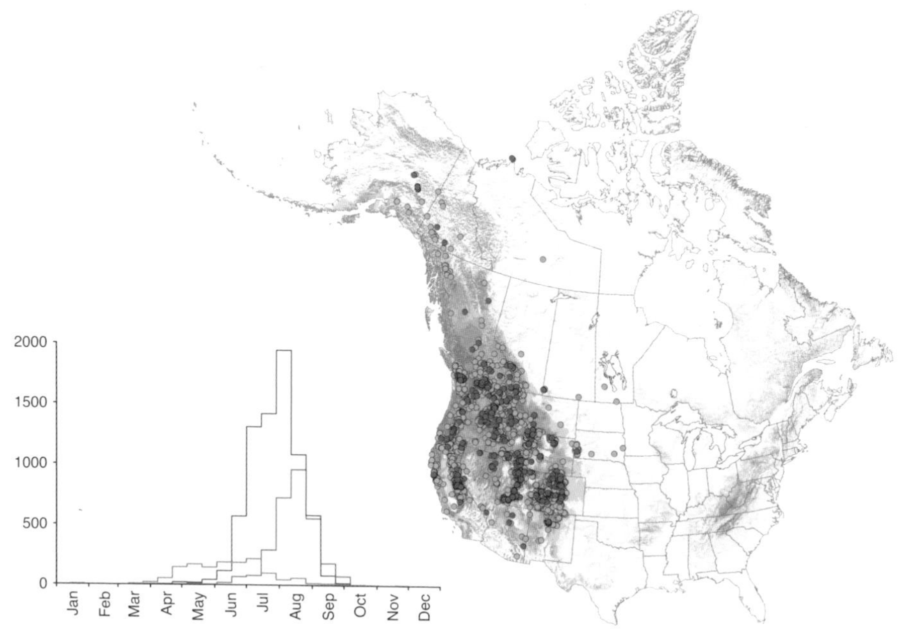
2000
1500
1000
500
0
Jan
Feb
Mar
Apr
May
Jun
Jul
Aug
Sep
Oct
Nov
Dec

HAND CHARACTERS Body size small: queen 15–19 mm (0.57–0.73 inch), worker 8–14 mm (0.31–0.55 inch). Hair short and even. Head length medium with the cheek (oculo-malar area) just shorter than broad (contrast *B. melanopygus, B. rufocinctus*), midleg basitarsus with the back far corner rounded, hindleg tibia outer surface flat without long hair but with long fringes at the sides, forming a pollen basket (corbicula). Hair *on the face yellow or white* (contrast *B. sylvicola*), sometimes with many black hairs intermixed on the upperside of the head and in the pale band at the front of the upperside of the thorax (contrast *B. huntii,* most *B. ternarius*), *pale band at the back of the upperside of the thorax interrupted in the middle broadly and entirely to the back edge with a black V shape* (contrast *B. sandersoni,* some *B. melanopygus*), sides of the thorax with black hair in at least the lower third, *hindleg with the pollen-basket fringes usually predominantly pale orangey brown* (contrast *B. ternarius, B. huntii, B. sylvicola, B. melanopygus, B. sandersoni*) or if metasomal T3 is black then sometimes the pollen-basket fringes are nearly black, T2 usually with an obvious patch of black hair in front in the middle (contrast some *B. melanopygus*), although sometimes T2–3 may show part or complete replacement of red with black, but if so then T4 is yellow and T5 is almost entirely black with yellow only at the extreme side and back fringes (contrast *B. melanopygus, B. sandersoni, B. sitkensis*). Male 8–13 mm (0.33–0.50 inch). Eye similar in size and shape to the eye of any female bumble bee (contrast *B. rufocinctus*). Antenna of medium length, flagellum 3× longer than the scape. Hair color pattern similar to the queen/worker, occasionally metasomal T1–5 mostly yellow with T3 and T6 predominantly black.

MICROSCOPIC CHARACTERS Queen/worker tarsi, especially the hindleg basitarsus, often brownish and paler than the tibia (contrast *B. sylvicola*). Male genitalia with the penis valve sickle-shaped, the back-curved "sickle" long and very narrow, at least 3× longer than broad, and less than half the breadth of the adjacent neck of the penis-valve head, the "sickle" scarcely flattened and about 2× broader than thick, almost spinelike and not expanded at the tip (contrast *B. sylvicola*), the penis-valve angle on the underside of the shaft and to the side slightly closer to the base than to the penis-valve head (contrast *B. sylvicola*), gonostylus as long as broad (contrast *B. sylvicola*), the inner (medial) edge strongly concave and curves basally inward beyond and below the inner furthest edge of the gonocoxa.

OCCURRENCE

RANGE AND STATUS Throughout the Mountain West from CA to AK, in the Western Desert of UT, NV, CO, AZ, NM, with scattered occurrences in the Tundra/Taiga of Canada and the Great Plains. From sea level to above 3,600 m. Common in most of its range.

HABITAT Open grassy prairies, urban parks and gardens, chaparral and shrub areas, mountain meadows.

EXAMPLE FOOD PLANTS *"Aster", Centaurea, Chrysothamnus, Cirsium, "Epilobium", Ericameria, Haplopappus, Helenium, Lupinus, Melilotus, Monardella, Penstemon, Ribes, Senecio, Solidago, Symphoricarpos.*

BEHAVIOR Nests usually underground, occasionally on the surface. Males patrol circuits in search of mates.

PARASITISM BY OTHER BEES Unknown.

BOMBUS CENTRALIS CRESSON, 1864
CENTRAL BUMBLE BEE

Worker *Bombus centralis*. GZ

Male *Bombus centralis*. MH

IDENTIFICATION

Western interior, medium long-tongued species. Most similar to some *B. flavifrons, B. mixtus,* and *B. frigidus* (see also *B. rufocinctus*). Evidence from DNA barcodes supports a close relationship between *B. centralis* and *B. flavifrons* but nonetheless supports them as separate species.

HAND CHARACTERS Body size small: queen 16–18 mm (0.64–0.70 inch), worker 10–13 mm (0.39–0.50 inch). Hair short and even. Head length medium with the cheek (oculo-malar area) slightly longer than broad (contrast *B. flavifrons, B. rufocinctus*), midleg basitarsus with the back far corner rounded, hindleg tibia outer surface flat without long hair but with long fringes at the sides, forming a pollen basket (corbicula). Hair of the upperside of the head and the upperside of the thorax at the front with the pale band yellow without many black hairs intermixed (contrast most *B. flavifrons, B. mixtus*), metasomal T1–2 either entirely yellow or with black at the front and in the middle, T3 orange (contrast *B. frigidus*), T5–6 mostly black. Male 12–14 mm (0.47–0.53 inch). Eye similar in size and shape to the eye of any female bumble bee (contrast *B. rufocinctus*). Antenna of medium length, flagellum 2.5–3× longer than the scape. Hair color pattern similar to the queen/worker, upperside of the thorax with the yellow band at the front with no black hairs intermixed at the front near the midline so that the black band between the wings is sharply distinct in the middle, but sometimes with yellow extensively intermixed in the black band, especially at the sides so that it may be less distinct at the sides, *upperside of the thorax with the yellow band at the back without black hairs except for a few at the front near the midline* (contrast *B. flavifrons*), metasomal *T3–4 red without any black hairs* (contrast *B. flavifrons*), T5–7 vary from mostly red to mostly black.

MICROSCOPIC CHARACTERS Male antennal segment A3 long, length nearly 2× its maximum breadth, almost as long as antennal segment A5. Genitalia with the penis valve sickle-shaped, the back-curved "sickle" short and very broad, less than 2× longer than the breadth of the broadest part and similar in breadth to the adjacent neck of the penis-valve head, the "sickle" flattened and broadest

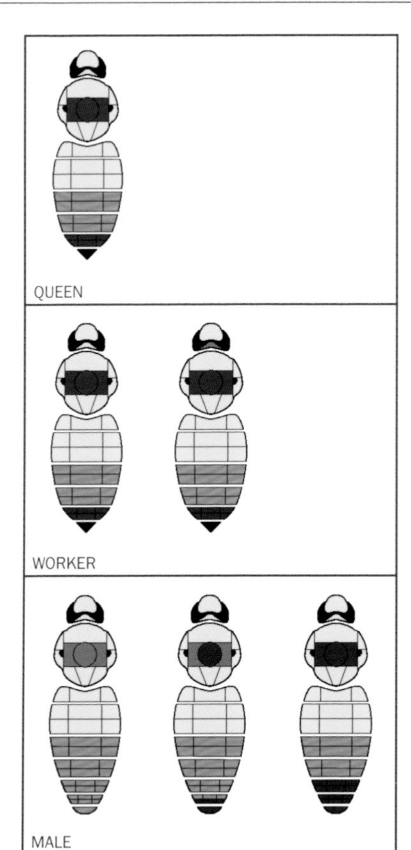

near the midpoint of its length, the penis-valve angle on the underside of the shaft and to the side located near the midpoint from its base to the penis-valve head, gonostylus with the inner (medial) edge nearly straight, the margin with a short indistinct parallel submarginal groove.

OCCURRENCE

RANGE AND STATUS Mountain West from southern CA to AK and adjacent Canada, southeast through Western Desert highlands of AZ, NM. From sea level to above 3,500 m. Moderately common in most of its range.

HABITAT Open grassy prairies and mountain meadows.

EXAMPLE FOOD PLANTS *Allium, Chrysothamnus, Cirsium, Ericameria, Monardella, Penstemon, Phacelia,*

BEHAVIOR Nests underground. Males patrol circuits in search of mates.

PARASITISM BY OTHER BEES Unknown.

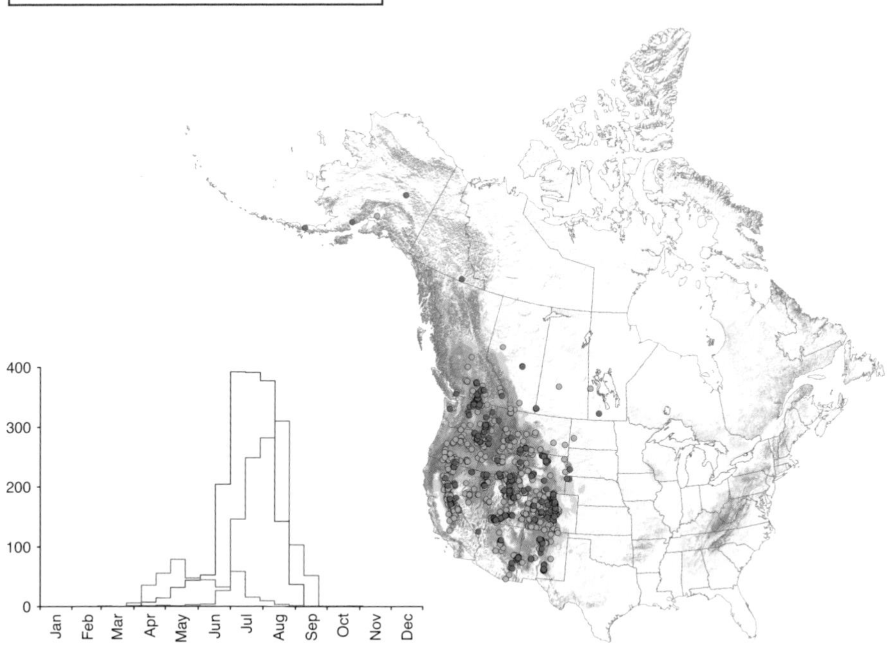

BOMBUS FLAVIFRONS CRESSON, 1863
YELLOW HEAD BUMBLE BEE (INCLUDING *DIMIDIATUS*)

ABOVE: **Male and Female *Bombus flavifrons*.** DK
RIGHT: **Male *Bombus flavifrons*.** LL
FAR RIGHT: **Worker *Bombus flavifrons*.** TS

IDENTIFICATION

Northern and western, long-tongued species. Most similar to *B. centralis*, *B. sitkensis*, and *B. mixtus* (see also *B. rufocinctus*). The oldest available name for this species is *B. pleuralis* Nylander, but the name *B. flavifrons* has been in common use for this species since 1950 and we know of only one publication using the name *B. pleuralis*. We suggest that, for stability, we should continue to use the name *B. flavifrons*. Evidence from DNA barcodes supports *B. centralis* and *B. flavifrons* as separate species. DNA barcodes also support as part of the same species individuals with metasomal T3–4 red (named *flavifrons*, from AK and BC south to UT) and those with T3–4 black (named *dimidiatus*, predominant in CA, OR, and WA).

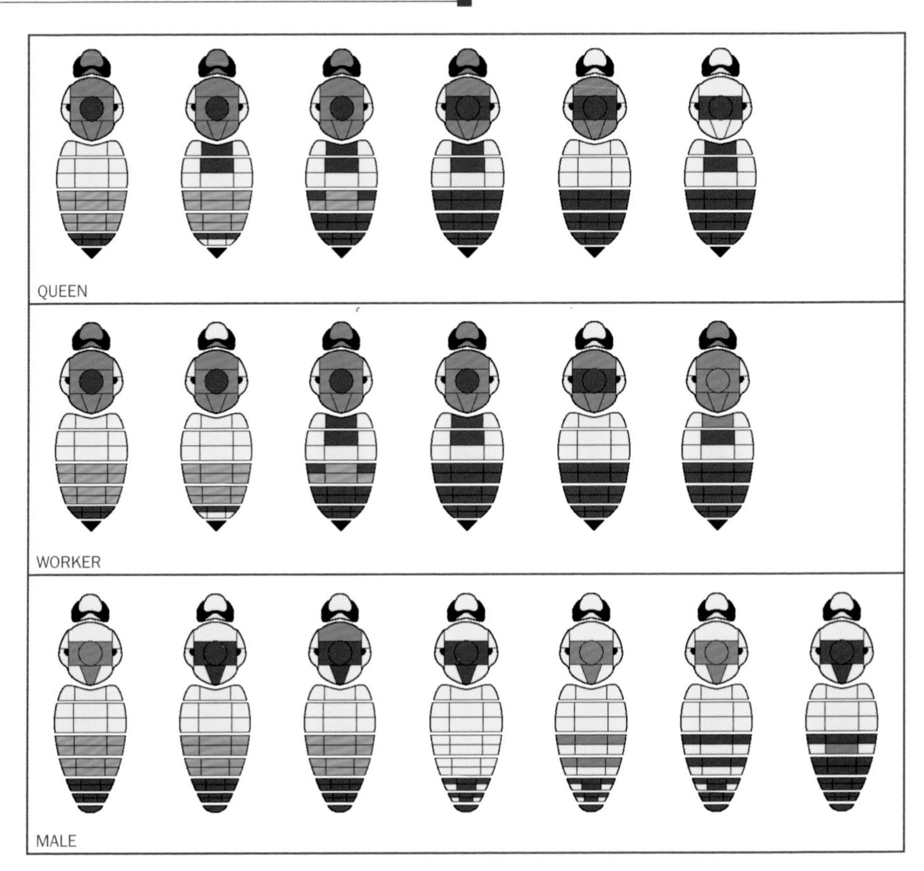
QUEEN
WORKER
MALE

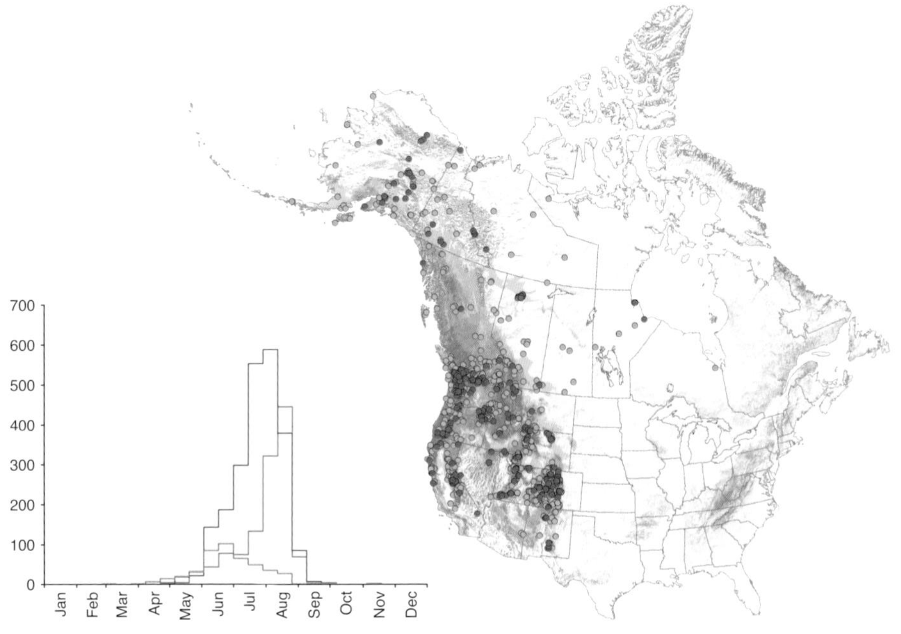
700
600
500
400
300
200
100
0
Jan
Feb
Mar
Apr
May
Jun
Jul
Aug
Sep
Oct
Nov
Dec

HAND CHARACTERS Body size small: queen 16–18 mm (0.62–0.69 inch), worker 10–13 mm (0.40–0.50 inch). Hair length medium (longer than *B. centralis*) and uneven. Head long with the cheek (oculo-malar area) *distinctly longer than broad* (contrast *B. centralis, B. sitkensis, B. rufocinctus*), midleg basitarsus with the back far corner rounded, hindleg tibia outer surface flat without long hair but with long fringes at the sides, forming a pollen basket (corbicula). Hair of the face, upperside of the head, and the upperside of the thorax at the front with the pale band yellow, usually with many black hairs intermixed (contrast *B. centralis*), at the back the pale band usually distinct with predominantly pale hair (contrast *B. sitkensis*), metasomal T1–2 with the yellow often interrupted in the middle by black especially broadly at the front (contrast *B. sitkensis*) or sometimes without black, T3–4 yellow at the sides and S3–5 predominantly yellow (contrast *B. vagans, B. sandersoni*), T5 black (contrast *B. sitkensis, B. mixtus*), sometimes with the fringe at the back yellow or rarely more extensively brownish yellow. Male 10–14 mm (0.41–0.56 inch). Eye similar in size and shape to the eye of any female bumble bee (contrast *B. rufocinctus*). Antenna of medium length, flagellum 2.5–3× longer than the scape. Hair color pattern similar to the queen/worker, but the upperside of the thorax at the front with the yellow band often with few or no black hairs intermixed near the midline, sometimes with more, making the black band between the wings indistinct, *upperside of the thorax with the yellow band at the back often with many scattered black hairs especially near the midline* (contrast *B. centralis*), metasomal *T3–4 almost always with at least a few black hairs intermixed near the front corners at the sides* (contrast *B. centralis, B. vandykei*), if T3–4 are red then T5 is sometimes also red, or if T3–4 are without red then they are often extensively yellow.
MICROSCOPIC CHARACTERS Female metasomal S6 with a strong ridge along the midline (contrast *B. sitkensis*). Male antennal segment A3 long, length nearly 2× its maximum breadth, almost as long as antennal segment A5. Genitalia with the penis valve sickle-shaped, the back-curved "sickle" short and very broad, less than 2× longer than the breadth of the broadest part and similar in breadth to the adjacent neck of the penis-valve head, the "sickle" flattened and broadest near the midpoint of its length (contrast *B. sitkensis*), the penis-valve angle on the underside of the shaft and to the side located near the midpoint from its base to the penis-valve head, gonostylus inner (medial) edge nearly straight, the margin thin with a short indistinct parallel submarginal groove.

OCCURRENCE

RANGE AND STATUS Mountain West from southern NM and AZ to AK, CA Northern Coast Ranges and southern Sierra Nevada, Tundra/Taiga of AK, and northwestern Canada to Hudson Bay. From sea level to above 2,700 m. Common throughout its range.
HABITAT Open grassy prairies and mountain meadows to transition and northern forest areas.
EXAMPLE FOOD PLANTS *Cirsium, "Epilobium", Heliomeris, Mentha, Penstemon, Vaccinium.*
BEHAVIOR Nests mostly underground, rarely on the surface. Males patrol circuits in search of mates.
PARASITISM BY OTHER BEES Host to *B. insularis*, confirmed breeding record.

BOMBUS SITKENSIS NYLANDER, 1848
SITKA BUMBLE BEE

LEFT: Female *Bombus sitkensis*. HB
ABOVE: Male *Bombus sitkensis*. AR

IDENTIFICATION

Western coastal, medium-tongued species. Most similar to *B. flavifrons, B. mixtus, B. bifarius,* and *B. melanopygus.*

HAND CHARACTERS Body size small: queen 15–20 mm (0.59–0.77 inch), worker 9–14 mm (0.36–0.55 inch). Hair long and shaggy and uneven. Head length medium with the cheek (oculo-malar area) *as long as broad* (contrast *B. flavifrons*), midleg basitarsus with the back far corner rounded, hindleg tibia outer surface flat without long hair but with long fringes at the sides, forming a pollen basket (corbicula). Hair of the upperside of the thorax with yellow but densely intermixed with black which predominates between the wings and behind the wings (contrast *B. mixtus, B. flavifrons*), metasomal T1 yellow, T2 with the yellow often narrowly interrupted by black *especially broadly at the back* or with black hairs intermixed (contrast *B. flavifrons*), T5 almost entirely pale brownish yellow (contrast *B. flavifrons, B. bifarius, B. melanopygus*). Male 9–14 mm (0.37–0.54 inch). Eye similar in size and shape to the eye of any female bumble bee. Antenna of medium length, flagellum 3× longer than the scape. Hair color pattern similar to the queen/worker, but the upperside of the thorax with the black band between the wings with many (more) yellow hairs intermixed, metasomal T3 usually with a pale fringe at the back or sometimes entirely yellow.

MICROSCOPIC CHARACTERS Female metasomal S6 evenly rounded or with only a weak ridge along the midline (contrast *B. flavifrons*). Male antennal segments A3–4 with the back edges without dense fringing patches of short hair (contrast *B. sandersoni*), at most with a very few short bristles, antennal segment A3 short, length less than 1.5× its maximum breadth, much shorter than antennal segment A5. Genitalia with the penis valve sickle-shaped, the back-curved "sickle" short and moderately broad, less than 2× longer than the breadth of the broadest part and similar in breadth to the adjacent neck of the penis-valve head, the "sickle" flattened and broadest near the tip (contrast *B. flavifrons*), the penis-valve angle on the underside of the shaft and to the side slightly closer to its base than to the penis-valve head, gonostylus inner (medial) edge nearly straight, the margin thick with a distinct long parallel submarginal groove.

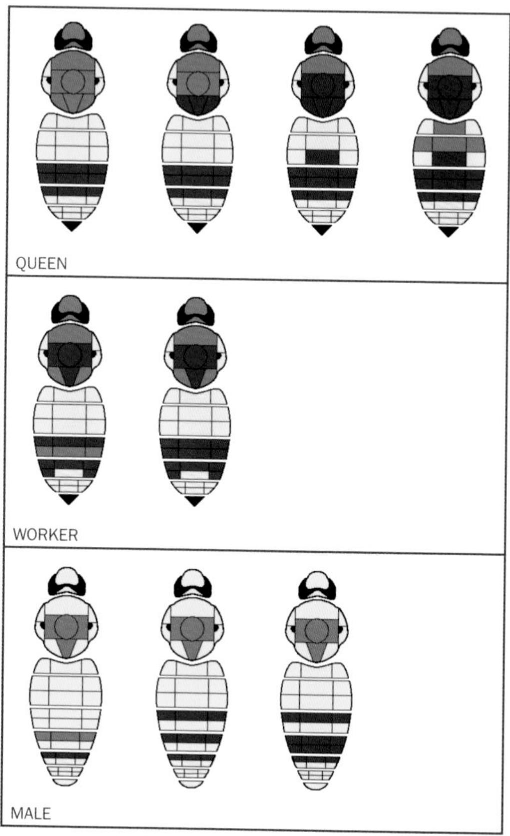

OCCURRENCE

RANGE AND STATUS CA Coastal Ranges north to AK, also some Mountain West populations, especially in CA Sierra Nevada, and in the Rocky Mountains of southern BC and northwestern ID, MT to WY.

HABITAT Open grassy prairies and mountain meadows.

EXAMPLE FOOD PLANTS *Ceanothus, Cirsium, Rhododendron, Ribes, Vaccinium.*

BEHAVIOR Nests underground. Males patrol circuits in search of mates.

PARASITISM BY OTHER BEES Unknown.

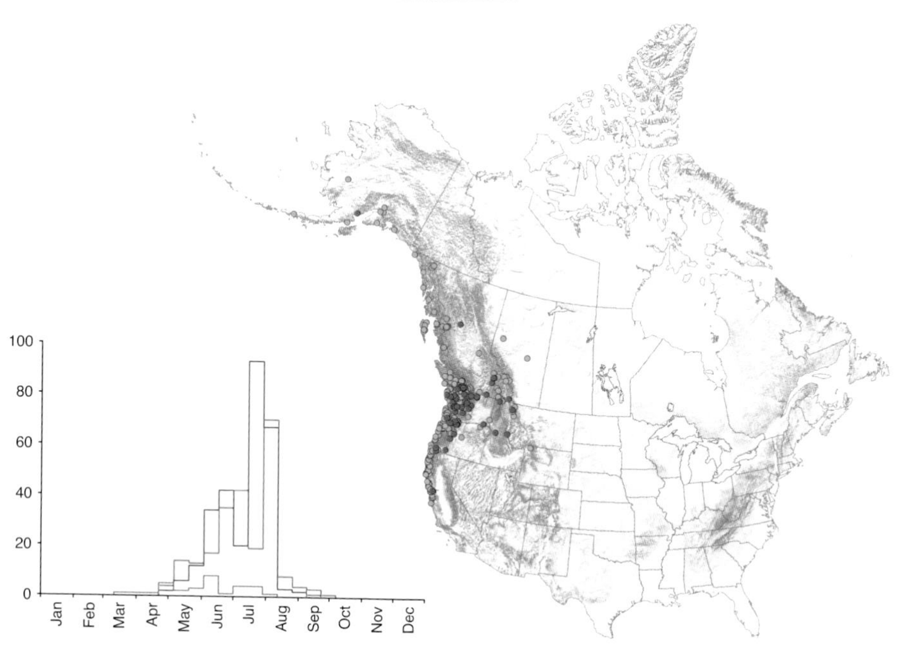

BOMBUS POLARIS CURTIS, 1835
POLAR BUMBLE BEE (INCLUDING *ARCTICUS* KIRBY, *GROENLANDICUS*, *DIABOLICUS*)

LEFT: **Queen *Bombus polaris*.** ROD
ABOVE: **Male *Bombus polaris*.** AR

IDENTIFICATION

Far northern, medium-tongued species. Most similar to *B. balteatus, B. neoboreus, B. hyperboreus, B. frigidus, B. flavifrons,* and *B. mixtus.* This species occurs in arctic Europe and Asia, where it is also very variable in color pattern. Evidence from DNA barcodes shows no support for more than one species among *B. polaris* with these color patterns in North America. The color pattern without yellow bands (named *diabolicus*) is known only from the NU mainland, but the color pattern with a black tail occurs as an infrequent variation throughout the population.

HAND CHARACTERS Body size medium: queen 18–22 mm (0.72–0.85 inch), worker 10–16 mm (0.39–0.61 inch). Hair very long and uneven. Head length medium with the *cheek* (oculo-malar area) *just longer than broad* (contrast *B. balteatus, B. hyperboreus*), midleg basitarsus with the back far corner just acute but rounded, hindleg tibia outer surface flat without long hair but with long fringes at the sides, forming a pollen basket (corbicula). Hair of the head usually entirely black, although the lightest workers with some yellow hairs intermixed on the upperside of the head, *sides of the thorax always with at least some black hair below and toward the back and often more extensive* (contrast *B. balteatus*), sides of the thorax may have yellow hair intermixed, and sometimes in workers the sides of the thorax are predominantly yellow although still with many black hairs intermixed, *for black-tailed specimens the upperside of the thorax front yellow band is often nearly twice the breadth of the back yellow band* measured along the body midline (contrast *B. hyperboreus*), the yellow very pale (contrast *B. hyperboreus*), metasomal *T3 often with some pale hair but this is concentrated in a central patch near the midline* while the sides of T3 are black or orange and without any trace of yellow lengthwise stripes at the edges (contrast *B. balteatus, B. neoboreus, B. hyperboreus*), the T3–4 banding pattern may be indistinct especially on workers because of extensive intermixing of the colored hair. Body shape rather short and broadly rounded. Male 14–16 mm (0.54–0.62 inch). Eye similar in size and shape to the eye of any

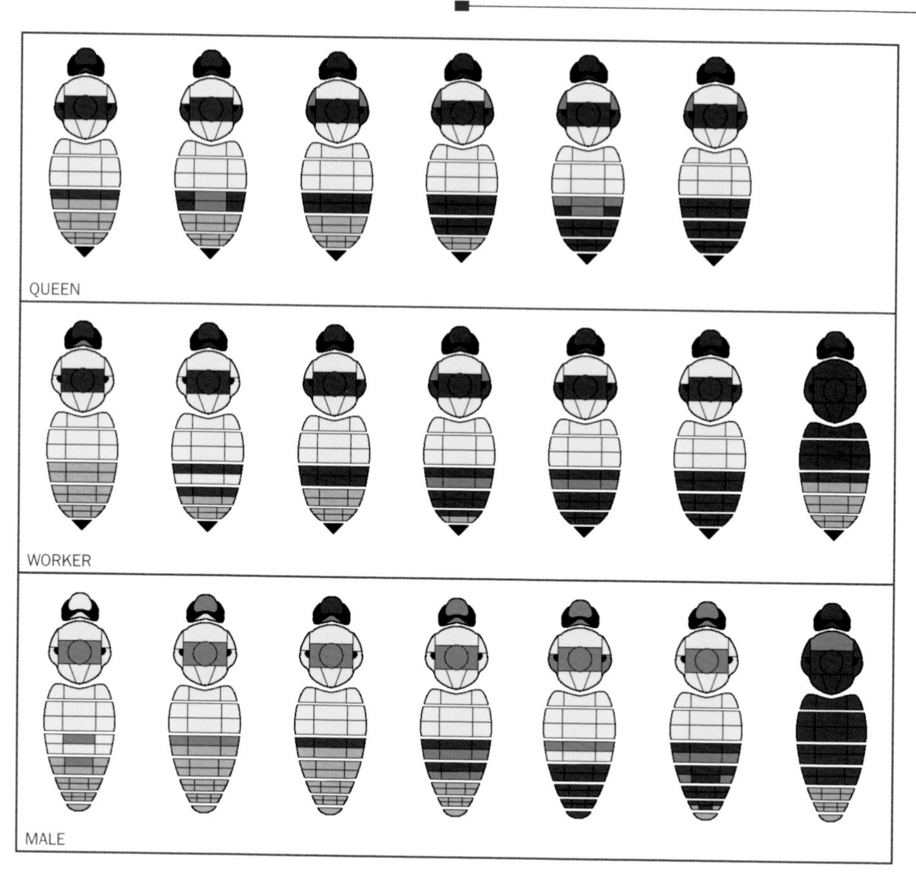
QUEEN
WORKER
MALE

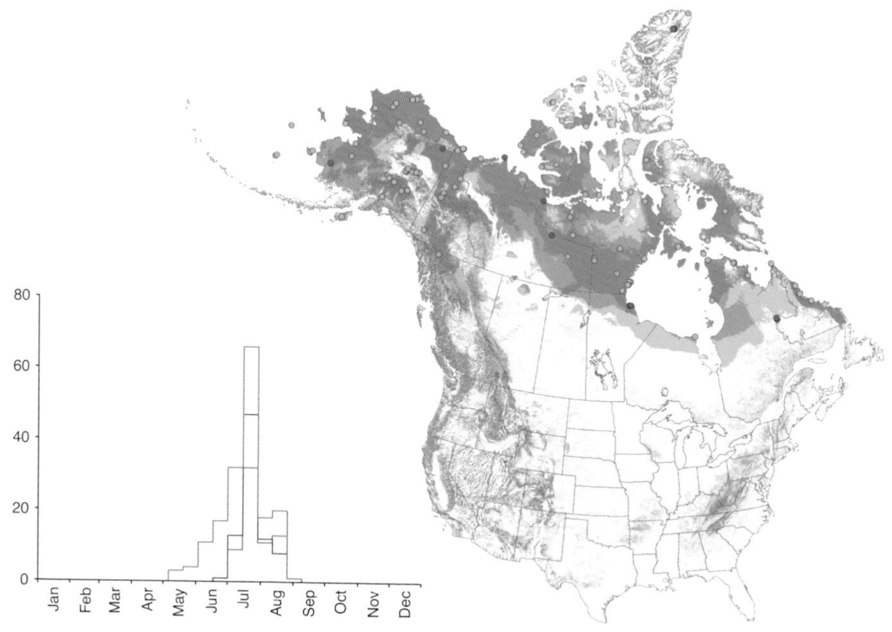
80
60
40
20
0
Jan
Feb
Mar
Apr
May
Jun
Jul
Aug
Sep
Oct
Nov
Dec

female bumble bee. Antenna long, flagellum nearly 4× longer than the scape. Hair color pattern similar to the queen/worker, sometimes predominantly black but often extensively pale, with patches of yellow on the face and on the upperside of the head, often with black intermixed, the upperside of the thorax between the wings and on metasomal T3–4 with black bands that often have pale hair extensively intermixed (contrast *B. balteatus, B. neoboreus*), T3–4 with fringes at the back weakly differentiated, at least T7 usually with some orange (contrast *B. hyperboreus*), S2–6 with yellow and black intermixed even if T3–6 are nearly black (contrast *B. hyperboreus*).
MICROSCOPIC CHARACTERS Queen/worker mandible with a deep notch in front of the back tooth (contrast *B. frigidus, B. flavifrons, B. mixtus*), clypeus from its central area toward the labrum with many medium and small pits or punctures (contrast *B. frigidus*), midleg basitarsus less than 3× longer than its greatest breadth (contrast *B. hyperboreus*), outer (corbicular) surface of the hindleg tibia matte and coarsely rough, with the surface texture strongly interrupting reflective highlights (contrast *B. hyperboreus*). Male hindleg tibia outer surface center smooth and shining, without short hairs, hairs of the back fringe longer than the maximum breadth of the tibia, hindleg basitarsus with the longest hairs in the back fringe more than 2× longer than the greatest breadth of the basitarsus (contrast *B. neoboreus*). Genitalia with the penis-valve head straight, the head with a strong triangular tooth on its outer edge below the tip by a distance approximately equal to the breadth of the penis-valve head (contrast *B. hyperboreus*), *gonostylus inner (medial) edge moderately concave, the submarginal groove indistinct and weakly defined, the front and back corners of the groove barely acute* (contrast *B. balteatus, B. neoboreus*).

OCCURRENCE

RANGE AND STATUS Tundra/Taiga region from AK north to Ellesmere Island and east to Baffin Island and northern QC, NF. Also in Europe and Asia.
HABITAT Tundra.
EXAMPLE FOOD PLANTS *Arnica, Lagotis, Pedicularis, Phyllodoce, Polemonium, Polygonum, Salix, Vaccinium.*
BEHAVIOR Nests underground. Males patrol circuits in search of mates.
PARASITISM BY OTHER BEES Host to *B. hyperboreus*, confirmed breeding record.

Queen *Bombus polaris*. BH

BOMBUS BALTEATUS DAHLBOM, 1832
HIGH COUNTRY BUMBLE BEE (INCLUDING *KIRBIELLUS*)

Worker (posed) *Bombus balteatus*. RP

Worker *Bombus balteatus*. DI

IDENTIFICATION

Far northern and western mountain, long-tongued species. Most similar to *B. polaris, B. neoboreus, B. hyperboreus, B. frigidus, B. flavifrons,* and *B. mixtus.* This species occurs in arctic Europe and Asia, where it is also very variable in color pattern. Evidence from DNA barcodes shows no support for more than one species among these bees in North America. The yellow-tailed color pattern is known from the NU mainland, the black-tailed pattern from the southern Rocky Mountains, and the pattern with more extensive black on the side of the thorax from Ellesmere Island.

HAND CHARACTERS Body size medium: queen 19–21 mm (0.76–0.85 inch), worker 11–19 mm (0.42–0.75 inch). Hair moderately long and even. Head long with the *cheek* (oculo-malar area) *much longer than broad* (contrast *B. polaris, B. frigidus*), midleg basitarsus with the back far corner just acute but rounded, hindleg tibia outer surface flat without long hair but with long fringes at the sides, forming a pollen basket (corbicula). Hair of the head black but often with a small tuft of yellow hairs at the antennal base and yellow hairs intermixed on the upperside of the head, *sides of the thorax yellow at least in the upper half and usually with only a few black hairs on the underside and toward the back* (contrast *B. hyperboreus*, most *B. polaris*) or only very rarely with black more extensively intermixed, metasomal *T3 usually with at least traces of yellow hair near the edges at the sides and often with a lengthwise stripe at the sides that continues around as a strongly differentiated fringe* along the back margin (contrast *B. polaris, B. neoboreus, B. hyperboreus*), T3–4 with the banding pattern clearly defined without extensive intermixing. Body shape rather elongated and pointed. Male 13–17 mm (0.52–0.66 inch). Eye similar in size and shape to the eye of any female bumble bee. Antenna long, flagellum nearly 4× longer than the scape. Hair color pattern similar to the queen/worker, but upperside of the thorax between the wings often with the black band with yellow extensively intermixed, metasomal S2–6 almost entirely bright yellow (contrast *B. polaris, B. neoboreus, B. hyperboreus*).

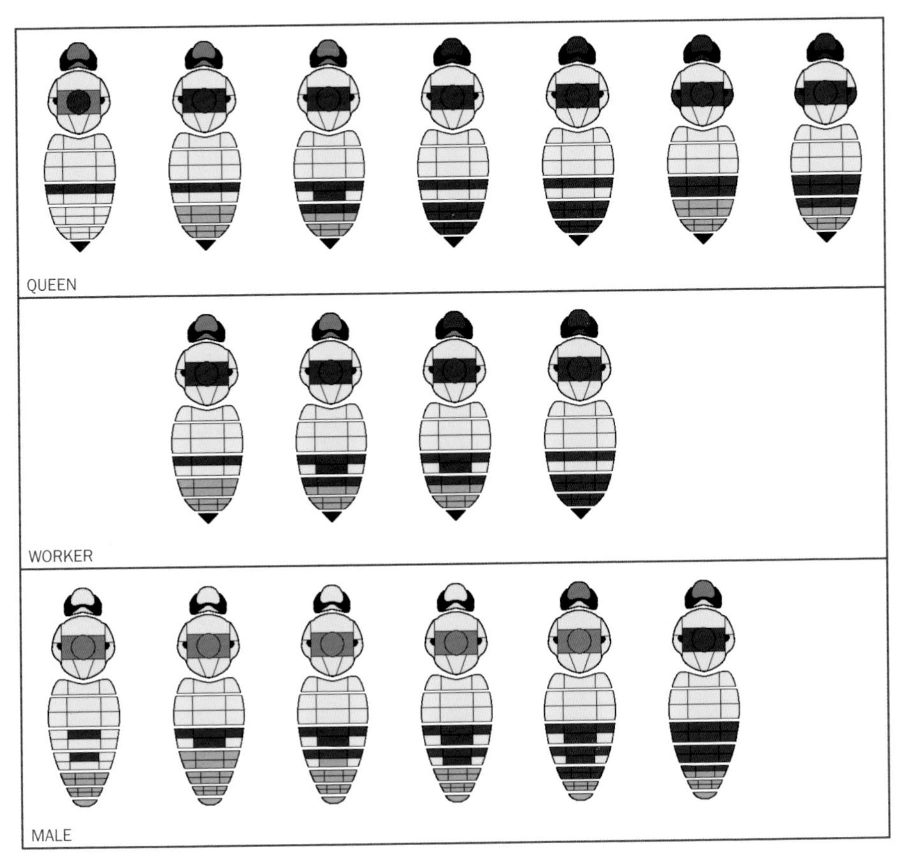
QUEEN
WORKER
MALE

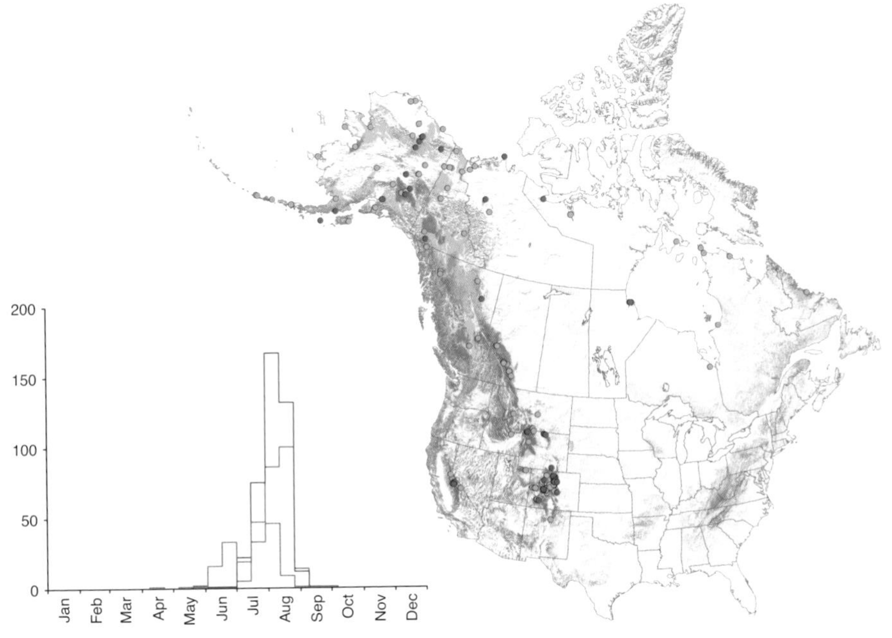
200
150
100
50
0
Jan
Feb
Mar
Apr
May
Jun
Jul
Aug
Sep
Oct
Nov
Dec

MICROSCOPIC CHARACTERS Queen/worker mandible with a deep notch in front of the back tooth (contrast *B. frigidus, B. flavifrons, B. mixtus*), midleg basitarsus less than 3× longer than its greatest breadth, outer (corbicular) surface of the hindleg tibia smooth and shiny (contrast *B. polaris*). Male hindleg tibia outer surface center smooth and shining, without short hairs, hairs of the back fringe longer than the maximum breadth of the tibia. Genitalia with the penis-valve head straight, the penis-valve head with a strong triangular tooth on its outer edge below the tip by a distance approximately equal to the breadth of the penis-valve head (contrast *B. hyperboreus*), *gonostylus inner (medial) edge strongly concave, the submarginal groove broad and sharply defined, the front and back corners of the groove strongly acute and pointed* (contrast *B. polaris*).

OCCURRENCE

RANGE AND STATUS Highest elevations of the Mountain West from CO and northern NM Rockies, CA Sierra Nevada and White Mountains, north through the Mountain West to AK, and Tundra/Taiga regions north to Ellesmere Island and east to northern QC, NF. Uncommon. Also in Europe and Asia.

HABITAT Open boreal areas mostly above the tree line.

EXAMPLE FOOD PLANTS *Castilleja, Chrysothamnus, Delphinium, "Epilobium", Mertensia, Penstemon.*

BEHAVIOR Nests underground. Males patrol circuits in search of mates.

PARASITISM BY OTHER BEES In Europe, host to *B. hyperboreus*, confirmed breeding record.

BOMBUS NEOBOREUS SLADEN, 1919
ACTIVE BUMBLE BEE (INCLUDING *STRENUUS*)

Female *Bombus neoboreus*. AR

Male *Bombus neoboreus*. AR

IDENTIFICATION

Far northern, long-tongued species. Most similar to *B. polaris, B. balteatus, B. hyperboreus, B. frigidus*, and *B. flavifrons*. Evidence from DNA barcodes appears to show that this species is not a single natural group (i.e., it is paraphyletic) without including *B. hyperboreus*. The color pattern with the hair of T3 black is known from the NT and NU mainland.

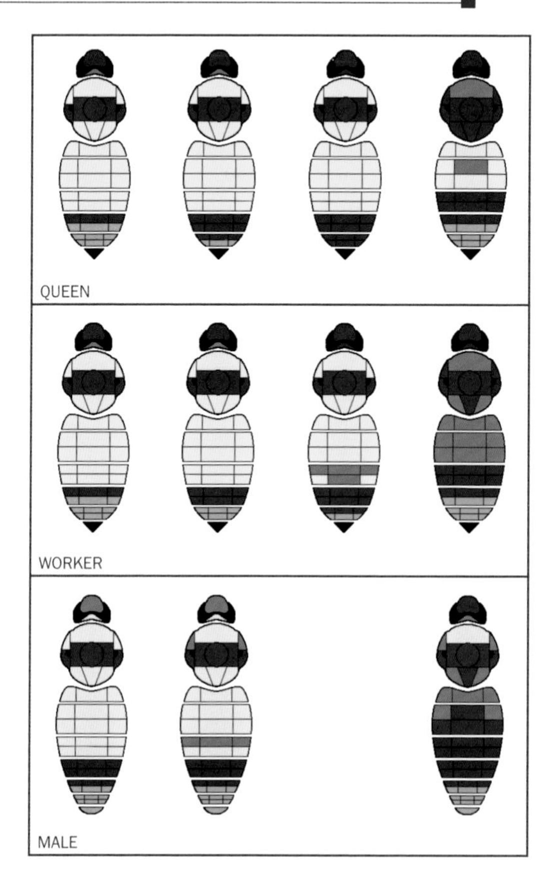
QUEEN
WORKER
MALE

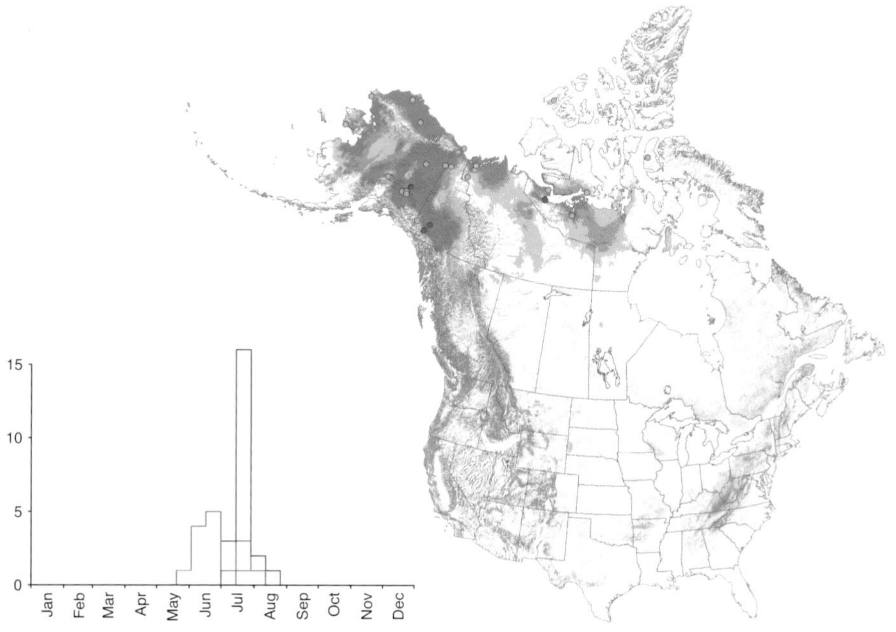
15
10
5
0
Jan
Feb
Mar
Apr
May
Jun
Jul
Aug
Sep
Oct
Nov
Dec

HAND CHARACTERS Body size large: queen 21–22 mm (0.83–0.87 inch), worker 10–13 mm (0.40–0.50 inch). Hair moderately long and even. Head long with the cheek (oculo-malar area) longer than broad (contrast *B. polaris, B. frigidus*), midleg basitarsus with the back far corner just acute but rounded, hindleg tibia outer surface flat without long hair but with long fringes at the sides, forming a pollen basket (corbicula). Hair of the face black but on the upperside of the head black with some yellow hairs intermixed, sides of the thorax yellow in at least the upper half or rarely entirely black, metasomal *T3 usually almost completely yellow* with only a few scattered black hairs (contrast *B. polaris, B. flavifrons, B. balteatus, B. hyperboreus, B. frigidus*) or if mostly black then the upperside of the thorax at the back (scutellum) black or at least yellow with many black hairs intermixed. Body shape rather elongated and rectangular. Male 17–18 mm (0.65–0.71 inch). Eye similar in size and shape to the eye of any female bumble bee. Antenna long, flagellum nearly 4× longer than the scape. Hair color pattern similar to the queen/worker, face black with a tuft of yellow hairs below the antenna base, upperside of the head yellow without black intermixed, the sides of the thorax yellow with some black at the back and below but with little intermixing, upperside of the thorax between the wings with the black band with few yellow hairs intermixed (contrast *B. polaris*), metasomal T3 usually predominantly yellow, or if black then the upperside of the thorax at the back (scutellum) yellow with many black hairs intermixed, T4 black with a small amount of yellow at the sides, T5 black with orange at the back (contrast most *B. polaris, B. balteatus*), T6 orange (contrast *B. hyperboreus*).
MICROSCOPIC CHARACTERS Queen/worker mandible with a deep notch in front of the back tooth (contrast *B. frigidus, B. flavifrons*), outer (corbicular) surface of the hindleg tibia smooth and shiny (contrast *B. polaris*). Male hindleg tibia outer surface center smooth and shining, without short hairs, hairs of the back fringe longer than the maximum breadth of the tibia, hindleg basitarsus with the longest hairs in the back fringe less than 2× longer than greatest breadth of the basitarsus (contrast *B. polaris*). Genitalia with the penis-valve head straight, *the head with only a small tooth on its outer edge below the tip by a distance approximately equal to the breadth of the penis-valve head* (contrast *B. polaris, B. hyperboreus*), gonostylus inner (medial) edge straight, the submarginal groove narrow and weakly defined, gonostylus outer edge reaching its furthest-back extremity where it joins the inner edge (contrast *B. hyperboreus*).

OCCURRENCE

RANGE AND STATUS Tundra/Taiga region from AK north to Prince Patrick Island and east to Baffin Island.
HABITAT Tundra.
EXAMPLE FOOD PLANTS Unknown.
BEHAVIOR Nests underground. Males patrol circuits in search of mates.
PARASITISM BY OTHER BEES Unknown.

BOMBUS HYPERBOREUS SCHÖNHERR, 1809
HIGH ARCTIC BUMBLE BEE (INCLUDING *ARCTICUS* QUENZEL)

LEFT: **Queen (Sweden) *Bombus hyperboreus*.** GH
ABOVE: **Male *Bombus hyperboreus*.** AR

IDENTIFICATION

Far northern, long-tongued species. Most similar to *B. polaris, B. balteatus, B. neoboreus,* and *B. flavifrons.* The oldest available name for this species is probably *B. arcticus* Quenzel (a previous attempt to suppress this name needs to be repeated appropriately), although the name *B. hyperboreus* has been in common use for this species since 1950 and applying the Quenzel name might lead to confusion with *B. arcticus* Kirby (= *B. polaris*). We suggest that, for stability, we should continue to use the name *B. hyperboreus.* This species occurs in arctic Europe and Asia, where one worker has been reported. All the supposed workers from North America we have seen so far are actually *B. polaris.* The males with the paler color pattern are known from NU and the darker males from NT.

HAND CHARACTERS Body size large: queen 21–24 mm (0.83–0.96 inch). Hair long and uneven. Head long, with the *cheek* (oculo-malar area) *much longer than broad* (contrast *B. polaris*), midleg basitarsus with the back far corner just acute but rounded, hindleg tibia outer surface flat without long hair but with long fringes at the sides, forming a pollen basket (corbicula). *Hair of the head and the sides of the thorax entirely black or with some yellow or brown hair on the upperside of the head at the back and occasionally with some yellow or brown hairs intermixed inconspicuously on the sides of the thorax* (contrast *B. polaris, B. balteatus*), upperside of the thorax with pale bands at the front and back usually of nearly equal breadth (measured along the body midline, contrast *B. polaris*), metasomal T3–5 entirely black, although the hairs sometimes with paler tips (contrast *B. neoboreus*), *pale hair often brownish rather than light yellow.* Body shape rather elongated and pointed. Male 17–19 mm (0.67–0.73 inch), very large. Eye similar in size and shape to the eye of any female bumble bee. Antenna long, flagellum nearly 4× longer than the scape. Hair color pattern similar to the queen, face black (contrast most *B. polaris*), upperside of the thorax between the wings with a black band usually without yellow hairs extensively intermixed (contrast *B. polaris, B. balteatus, B. neoboreus*) although sometimes these form a minority, metasomal T3 black or sometimes with yellow hairs especially at the sides and back, *T4–7 and S2–6 usually black* (contrast *B. polaris, B. balteatus, B. neoboreus*) or sometimes T4 more yellow and very rarely T6–7 with orange.

MICROSCOPIC CHARACTERS Queen mandible with a deep notch in front of the back tooth (contrast *B. flavifrons*), midleg basitarsus more than 3× longer than its greatest breadth (contrast *B. polaris*), outer (corbicular) surface of the hindleg tibia shiny, moderately rough, with the surface texture weakly interrupting the reflective highlights (contrast *B. polaris*). Male hindleg tibia with the outer surface center smooth and shining, without short hairs, hairs of the back fringe longer than the maximum breadth of the tibia. Genitalia with the penis head straight, *the head without*

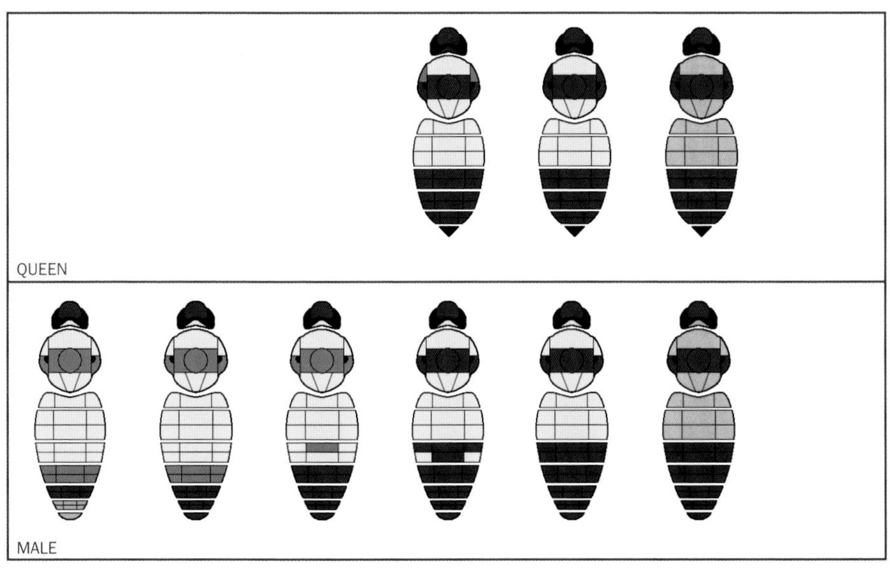

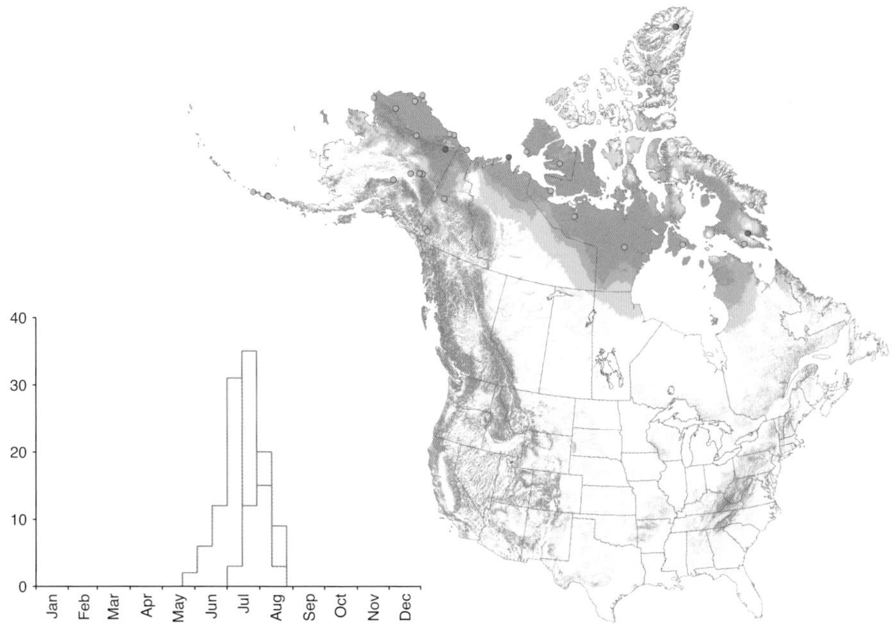

a tooth on its outer edge below the tip by a distance approximately equal to the breadth of the penis-valve head (contrast *B. polaris, B. balteatus, B. neoboreus*) or at most with a very small point that does not break the outer curve of the penis-valve head, gonostylus inner (medial) edge nearly straight with a slight convexity at two-thirds the distance from the near to the far end, the submarginal groove fused and weakly defined, gonostylus with the outer edge reaching its furthest-back extremity before curving forward to join the inner edge (contrast *B. neoboreus*).

OCCURRENCE

RANGE AND STATUS Tundra/Taiga region from AK north to Ellesmere Island and east to Baffin Island. Also in Europe and Asia.
HABITAT Tundra.
EXAMPLE FOOD PLANTS *Pedicularis, Salix, Saxifraga.*
BEHAVIOR Social parasite. Males patrol circuits in search of mates.
PARASITISM OF OTHER BEES This species is recorded as usually breeding as a parasite in colonies of *B. polaris*, although in Europe there is one record of it breeding in colonies of *B. balteatus*.

LEFT: **Female *Bombus terricola*.** LR
ABOVE: **Male *Bombus terricola*.** LR

LEFT AND ABOVE: **Workers *Bombus terricola*.** LR

IDENTIFICATION

Eastern and northern, short-tongued species. Most similar to *B. occidentalis* and *B. cryptarum* (see also *B. pensylvanicus, B. auricomus,* and *B. nevadensis*). Evidence from DNA barcodes supports this as a species separate from *B. occidentalis,* which it more or less replaces in the east.
HAND CHARACTERS Body size medium: queen 19–21 mm (0.73–0.84 inch), worker 10–15 mm (0.39–0.57 inch). Hair short and even. Head short with the cheek (oculo-malar area) just shorter than broad (contrast *B. pensylvanicus, B. auricomus, B. nevadensis*), midleg basitarsus with the back far corner rounded, hindleg tibia outer surface flat without long hair but with long fringes at the sides, forming a pollen basket (corbicula). Hair of the head black or with a minority of

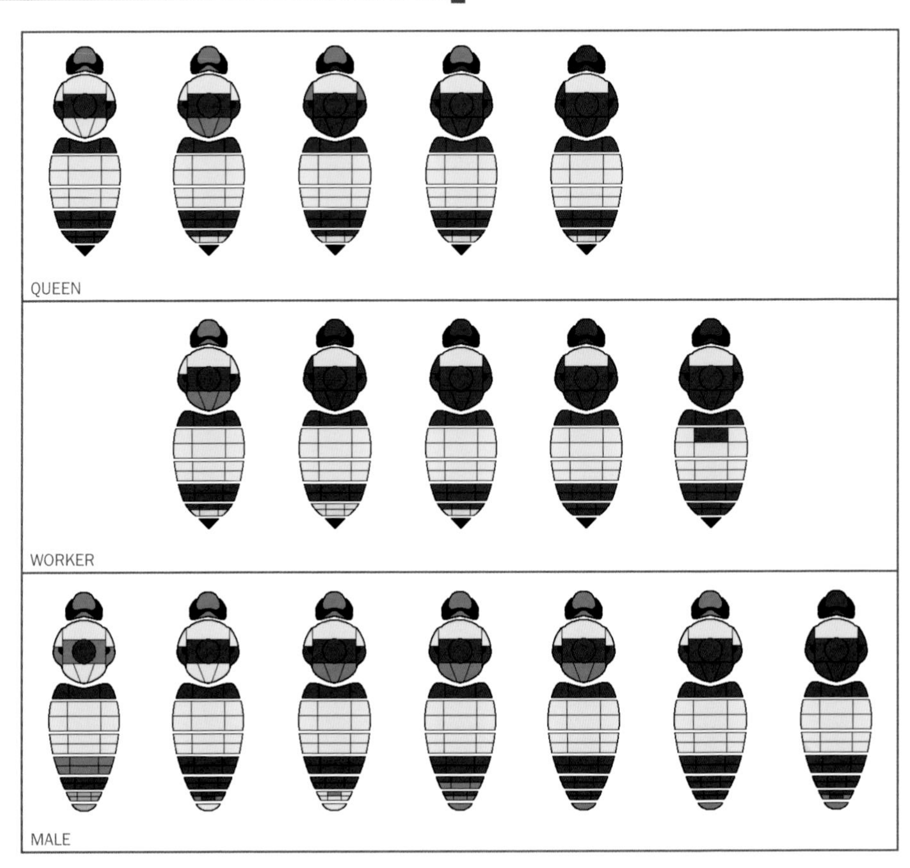
QUEEN
WORKER
MALE

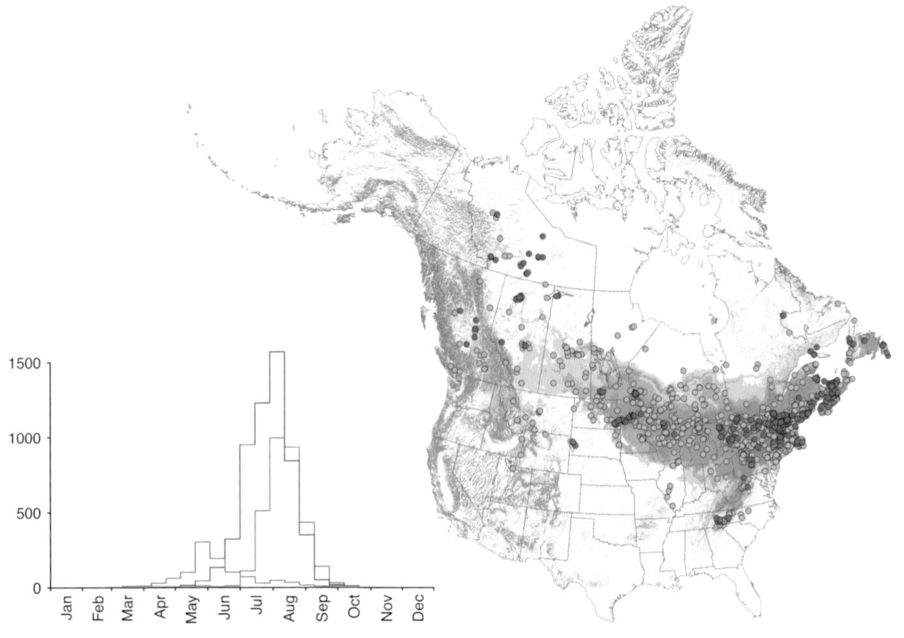
1500
1000
500
0
Jan
Feb
Mar
Apr
May
Jun
Jul
Aug
Sep
Oct
Nov
Dec

short pale hairs intermixed, metasomal *T2 at the front usually yellow without black or with only a narrow fringe along the front margin, but if T2 is more extensively black then T4–5 are also predominantly black* (contrast *B. occidentalis*), T3 usually yellow and T5 black or yellow-brown (contrast most *B. cryptarum*). Wings slightly brown (contrast *B. cryptarum*). Male 13–15 mm (0.50–0.60 inch). Eye similar in size and shape to the eye of any female bumble bee (contrast *B. auricomus, B. nevadensis*). Antenna short, flagellum just over 2× longer than the scape (contrast *B. pensylvanicus*). Hair color pattern similar to the queen/worker.

MICROSCOPIC CHARACTERS Queen/worker clypeus strongly swollen in the upper half, *clypeus central area predominantly smooth with scattered small pits or punctures* (contrast *B. cryptarum*), mandible with a deep notch in front of the back tooth, ocelli small and located on a line between the back edges of the eyes, hindleg basitarsus with the back edge strongly and evenly arched (contrast *B. pensylvanicus, B. auricomus, B. nevadensis*). Male genitalia with the penis-valve head greatly broadened from the upperside to the underside and flared outward to form (half of) a broad funnel (contrast *B. pensylvanicus, B. auricomus, B. nevadensis*).

OCCURRENCE

RANGE AND STATUS NL and Eastern Temperate Forest and Boreal Forest regions, south in a narrow band at higher elevations along the Appalachian Mountains, west through ND and the Canadian Great Plains, to the Tundra/Taiga of Canada and the Mountain West, especially in BC. Previously common throughout the northeast US and most of Canada, this species has declined in parts of its range.

HABITAT Close to or within wooded areas and wetlands.

EXAMPLE FOOD PLANTS *Crocus, Eupatorium, Linaria, Melilotus, Monarda, Ribes, Rosa, Rubus, Spiraea, Taraxacum, Vaccinium, Vicia.*

BEHAVIOR Nests underground. A frequent nectar robber of long-corolla flowers. Males patrol circuits in search of mates.

PARASITISM BY OTHER BEES Host to *B. bohemicus (=ashtoni)*, confirmed breeding record. It is likely that this species is also a host to *B. suckleyi* and *B. insularis.*

BOMBUS OCCIDENTALIS GREENE, 1858
WESTERN BUMBLE BEE (INCLUDING *MCKAYI*)

ABOVE: **Female *Bombus occidentalis*.** AF
LEFT: **Male *Bombus occidentalis*.** AF

IDENTIFICATION

Western, short-tongued species. Most similar to *B. terricola, B. cryptarum,* and *B. franklini* (see also *B. vosnesenskii, B. caliginosus, B. vandykei, B. fervidus, B. pensylvanicus, B. auricomus, B. nevadensis, B. insularis,* and *B. flavidus*). Evidence from DNA barcodes supports this as a species separate from *B. terricola.* This species more or less replaces *B. terricola* in most of the west, other than in southern BC. DNA barcode evidence also supports two subgroups with low

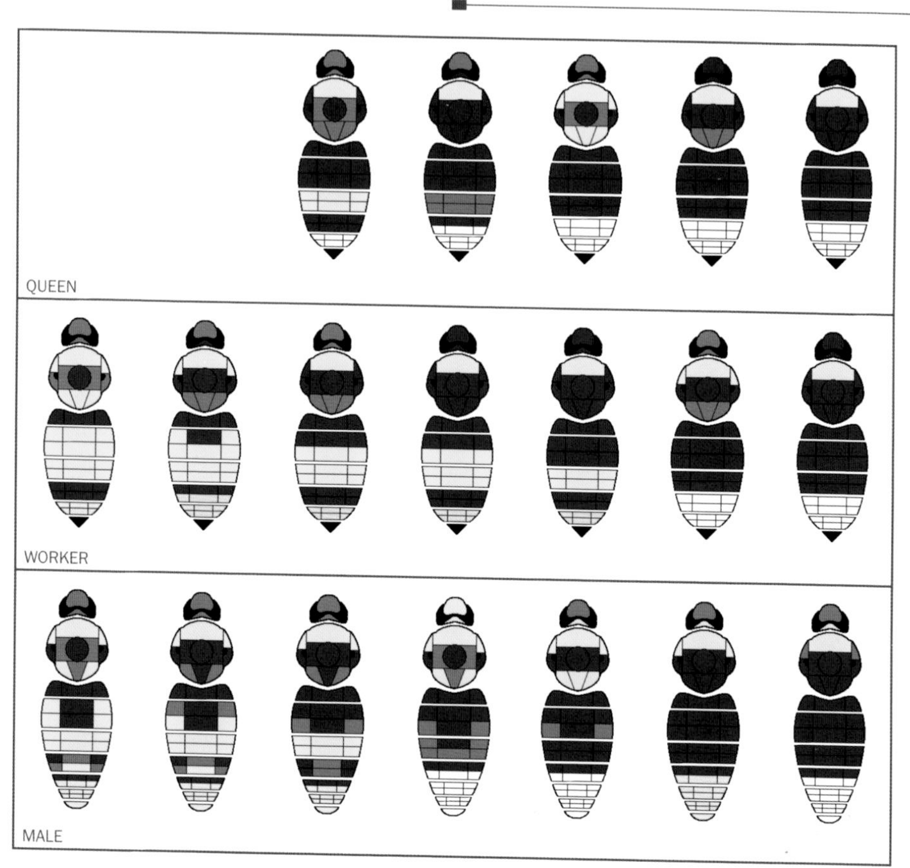
QUEEN
WORKER
MALE

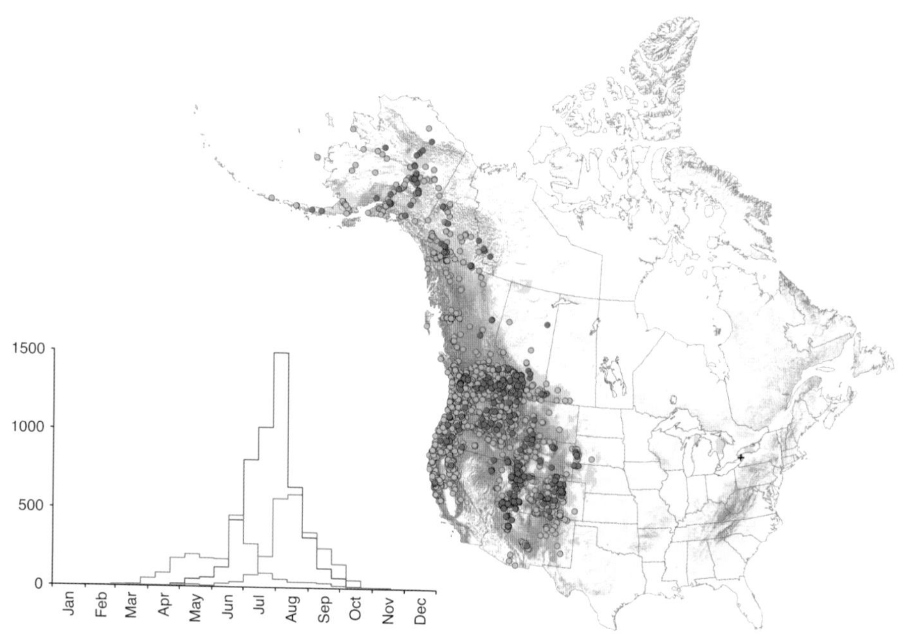
1500
1000
500
0
Jan
Feb
Mar
Apr
May
Jun
Jul
Aug
Sep
Oct
Nov
Dec

divergence: (1) a northern long-haired group (named *mckayi*, from AK and YT); and (2) a southern short-haired group (named *occidentalis*, from CA, CO, ID, BC, AB, SK, ON). The darkest color patterns are confined to the Coastal Range of CA (and northward into BC), whereas the palest color patterns occur in the Midwest.

HAND CHARACTERS Body size medium: queen 20–21 mm (0.77–0.84 inch), worker 9–15 mm (0.36–0.59 inch). Hair moderately short and even. Head short with the cheek (oculo-malar area) just shorter than broad (contrast *B. vosnesenskii, B. caliginosus, B. vandykei, B. fervidus, B. pensylvanicus, B. auricomus, B. nevadensis*), midleg basitarsus with the back far corner rounded, hindleg tibia outer surface flat without long hair but with long fringes at the sides, forming a pollen basket (corbicula). *Hair of the upperside of the thorax between the wings with at least a large black central spot and often a black band between the wings* (contrast *B. franklini*), metasomal *T2 usually black at the front, but if T2 is yellow at the front with just a few black hairs along the front margin then the head and thorax are also predominantly yellow* (contrast *B. terricola, B. cryptarum*), T3 often partly black but usually with at least a few scattered yellow hairs (contrast *B. terricola, B. cryptarum, B. franklini*), or *if* T2–3 are entirely black *then* T5–6 are extensively white or yellow-orange. Even the darkest queens and workers with a white tail have some grayish hair on the face and on the upperside of the head. Male 12–16 mm (0.47–0.61 inch). Eye similar in size and shape to the eye of any female bumble bee (contrast *B. auricomus, B. nevadensis*). Antenna short, flagellum just over 2× longer than the scape (contrast *B. vosnesenskii, B. caliginosus, B. vandykei, B. fervidus, B. pensylvanicus*). Hair color pattern similar to the queen/worker.

MICROSCOPIC CHARACTERS Queen/worker clypeus strongly swollen in the upper half, the *clypeus central area predominantly smooth with scattered small pits or punctures* (contrast *B. cryptarum*), mandible with a deep notch in front of the back tooth, ocelli small and located on a line between the back edges of the eyes, hindleg basitarsus with the back edge strongly and evenly arched (contrast *B. vosnesenskii, B. caliginosus, B. vandykei, B. fervidus, B. pensylvanicus, B. auricomus, B. nevadensis, B. insularis, B. flavidus*). Male genitalia with the penis-valve head greatly broadened from the upperside to the underside and flared outward to form (half of) a broad funnel (contrast *B. vosnesenskii, B. caliginosus, B. vandykei, B. fervidus, B. pensylvanicus, B. auricomus, B. nevadensis, B. insularis, B. flavidus*).

OCCURRENCE

RANGE AND STATUS Mountain West south to AZ, NM, Mediterranean CA, Pacific Coast north to AK Tundra/Taiga region, northwestern Great Plains east to southern SK. From sea level to above 2,000 m. Formerly common throughout much of its range, but populations from central CA to southern BC and west of the Sierra-Cascade Ranges have declined sharply since the late 1990s.

HABITAT Open grassy areas, urban parks and gardens, chaparral and shrub areas, mountain meadows.

EXAMPLE FOOD PLANTS *Ceanothus, Centaurea, Chrysothamnus, Cirsium, Geranium, Grindellia, Lupinus, Melilotus, Monardella, Rubus, Solidago, Trifolium.*

BEHAVIOR Nests usually underground. A frequent nectar robber of long-corolla flowers. Males patrol circuits in search of mates.

PARASITISM BY OTHER BEES Host to *B. suckleyi*, confirmed breeding record. It is likely that this species is also a host to *B. bohemicus (=ashtoni), B. insularis*, and *B. flavidus (=fernaldae).*

BOMBUS CRYPTARUM (FABRICIUS, 1775)
CRYPTIC BUMBLE BEE (INCLUDING *MODERATUS*)

ABOVE: **Female *Bombus cryptarum*.** SCA

LEFT: **Male *Bombus cryptarum*.** HP

IDENTIFICATION

Northwestern, short-tongued species. Most similar to *B. occidentalis* and *B. terricola* (see also *B. jonellus* and *B. auricomus*). Evidence from DNA barcodes supports this population as part of a species widespread in Europe and Asia. The very darkest color pattern of the worker is known from just one individual from near Hudson Bay.

HAND CHARACTERS Body size medium: queen 19–21 mm (0.75–0.82 inch), worker 13–17 mm (0.49–0.65 inch). Hair length medium. Head short with the cheek (oculo-malar area) just shorter than broad (contrast *B. jonellus, B. auricomus*), midleg basitarsus with the back far corner rounded, hindleg tibia outer surface flat without long hair but with long fringes at the sides, forming a pollen basket (corbicula). Hair of the head black, metasomal *T2 pale yellow with a narrow black fringe at the back* (or rarely extensively black) *and T3 black* (contrast *B. terricola, B. occidentalis*). Some workers with more or less pale hairs intermixed on the upperside of the thorax behind the wings (scutellum). Male 14–17 mm (0.54–0.66 inch). Eye similar in size and shape to the eye of any female bumble bee (contrast *B. auricomus*). Antenna short, flagellum just over 2× longer than the scape (contrast *B. jonellus*). Hair color pattern similar to the queen/worker, although sometimes the upperside of the thorax at the back (scutellum) and metasomal T1 yellow.

MICROSCOPIC CHARACTERS Queen/worker clypeus strongly swollen in the upper half, *clypeus central area coarsely sculptured with medium and large pits or punctures* (contrast *B. occidentalis, B. terricola*), mandible with a deep notch in front of the back tooth, ocelli small and located on

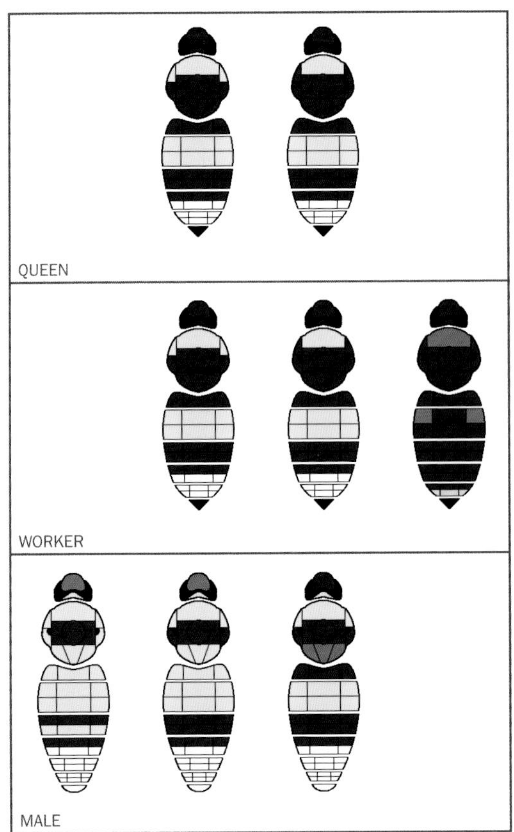

a line between the back edges of the eyes, hindleg basitarsus with the back edge strongly and evenly arched. Male genitalia with the penis-valve head greatly broadened from the upperside to the underside and flared outward to form (half of) a broad funnel (contrast *B. jonellus, B. auricomus*).

OCCURRENCE

RANGE AND STATUS Tundra/Taiga region from the Mountain West of AK and northwestern Canada to Hudson Bay. Range expanding south and eastward into AB over the last 20 years. Also in Europe and Asia.
HABITAT Tundra/Taiga.
EXAMPLE FOOD PLANTS *"Epilobium", Melilotus, Potentilla, Senecio.*
BEHAVIOR Nests underground. Males patrol circuits in search of mates.
PARASITISM BY OTHER BEES Unknown.

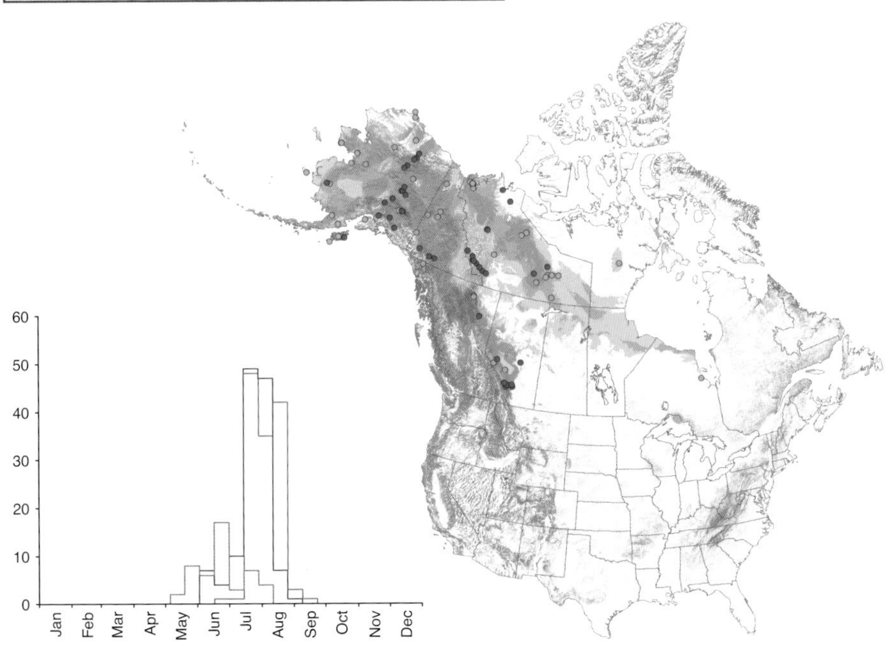

BOMBUS FRANKLINI (FRISON, 1921)
FRANKLIN BUMBLE BEE

ABOVE: **Female *Bombus franklini*.** RTH
RIGHT: **Female *Bombus franklini*.** AR
FAR RIGHT: **Male *Bombus franklini*.** AR

IDENTIFICATION

Western coastal, short-tongued species. Most similar to *B. occidentalis* (see also *B. vosnesenskii, B. caliginosus, B. vandykei, B. fervidus, B. insularis,* and *B. flavidus*). This species may now be extinct.

HAND CHARACTERS Body size large: queen 22–24 mm (0.86–0.95 inch), worker 10–17 mm (0.40–0.65 inch). Hair very short and even. Head short with the cheek (oculo-malar area) just shorter than broad (contrast *B. vosnesenskii, B. caliginosus, B. vandykei, B. fervidus*), midleg basitarsus with the back far corner rounded, hindleg tibia outer surface flat without long hair but with long fringes at the sides, forming a pollen basket (corbicula). *Hair of the upperside of the thorax (almost the entire scutum) yellow at the front and sides, between the wings with a small patch of black hair* (at the back and in the middle on the scutum) *and behind the wings black* (scutellum), giving the yellow a characteristic inverted U-shape (contrast *B. occidentalis*). Male 13–16 mm (0.50–0.64 inch). Eye similar in size and shape to the eye of any female bumble bee. Antenna short, flagellum just over 2× longer than the scape (contrast *B. vosnesenskii, B. caliginosus, B. vandykei, B. fervidus*). Hair color pattern similar to the queen/worker.

MICROSCOPIC CHARACTERS Queen/worker clypeus strongly swollen in the upper half, mandible with a deep notch in front of the back tooth, ocelli small and located on a line between the back edges

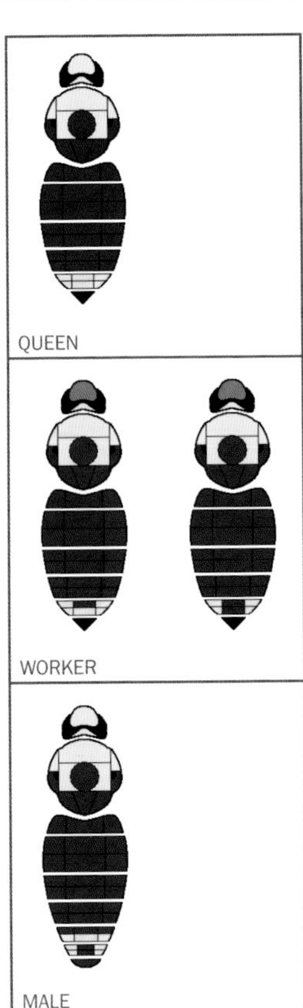

of the eyes, hindleg basitarsus with the back edge strongly and evenly arched (contrast *B. vosnesenskii, B. caliginosus, B. vandykei, B. fervidus, B. insularis, B. flavidus*). Male genitalia with the penis-valve head greatly broadened from the upperside to the underside and flared outward to form (half of) a broad funnel (contrast *B. vosnesenskii, B. caliginosus, B. vandykei, B. fervidus, B. insularis, B. flavidus*).

OCCURRENCE

RANGE AND STATUS Southern OR to northern CA. This species has always had one of the narrowest recorded geographic distributions among bumble bees worldwide, of about 34,450 km^2 (13,300 sq miles). Within this area it occurs at elevations from about 150 m to more than 2,000 m (500 ft to >6,700 ft). Always uncommon, this species declined sharply after 1999. It was last seen in 2006 and there are concerns that the species may be extinct.

HABITAT Open grassy coastal prairies, Coast Range meadows.

EXAMPLE FOOD PLANTS *Agastache, Centaurea, Ceonothus, Eriogonum, Eschscholzia, Lupinus, Monardella, Vicia.*

BEHAVIOR Nests underground. Males patrol circuits in search of mates.

PARASITISM BY OTHER BEES Unknown.

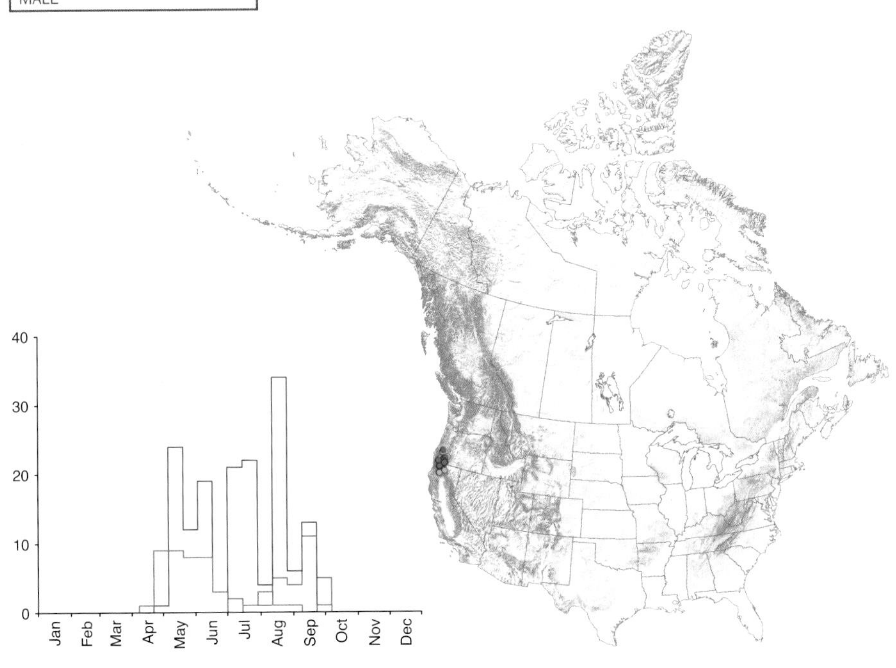

BOMBUS AFFINIS CRESSON, 1863
RUSTY-PATCHED BUMBLE BEE

LEFT: **Queen *Bombus affinis*.** SCAR
BELOW LEFT: **Male *Bombus affinis*.** GZ
BELOW RIGHT: **Worker *Bombus affinis*.** CS

IDENTIFICATION

Eastern, short-tongued species. Most similar to *B. griseocollis, B. fraternus, B. morrisoni,* and *B. rufocinctus* (see also *B. perplexus, B. vagans, B. bimaculatus, B. impatiens,* and *B. sandersoni*).
HAND CHARACTERS Body size large: queen 19–23 mm (0.75–0.92 inch), worker 9–16 mm (0.37–0.64 inch). Hair moderately long (contrast *B. griseocollis, B. fraternus, B. morrisoni, B. rufocinctus*) and even. Head short with the cheek (oculo-malar area) just shorter than broad (contrast *B. perplexus, B. vagans, B. bimaculatus, B. impatiens, B. sandersoni, B. rufocinctus*), midleg basitarsus with the back far corner rounded, hindleg tibia outer surface flat without long hair but with long fringes at the sides, forming a pollen basket (corbicula). Hair of the face black, *hair of the upperside of the head black* (contrast *B. morrisoni, B. rufocinctus*) or with only a few yellow hairs intermixed, thorax predominantly yellow including the sides (contrast *B. fraternus, B. morrisoni*), or upperside of the thorax with a black spot or band between the wings, metasomal T1–2 yellow or the *worker T2 often extensively brownish with more yellow at the edges* (contrast *B. fraternus, B. morrisoni, B. rufocinctus*), T3–5 black (contrast *B. morrisoni*). Rarely the hair of

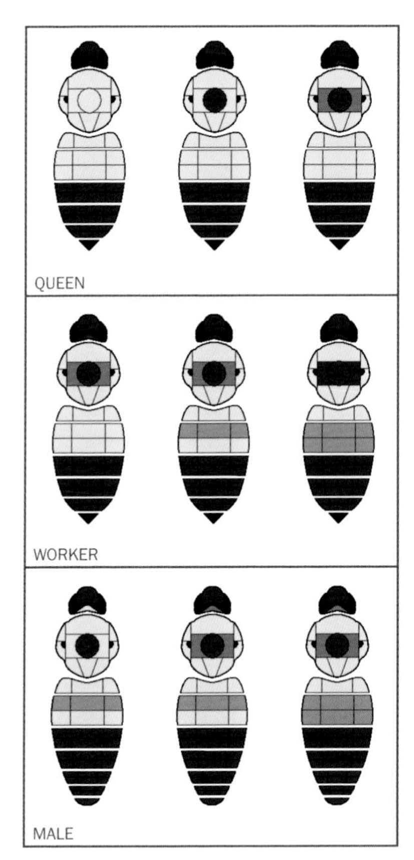

T2–5 orange-red at the back, although this may be from damage at the pupal stage. Male 14–17 mm (0.55–0.66 inch). Eye similar in size and shape to the eye of any female bumble bee. Antenna short, flagellum just over 2× longer than the scape (contrast *B. perplexus, B. vagans, B. bimaculatus, B. impatiens, B. sandersoni*). Hair color pattern similar to the queen/worker, but metasomal T2 usually brownish (one individual has extensive orange-yellow hair on T4).

MICROSCOPIC CHARACTERS Queen/worker mandible with a deep notch in front of the back tooth (contrast *B. griseocollis, B. fraternus, B. morrisoni, B. rufocinctus*), clypeus strongly swollen in its upper part, ocelli small and located on a line between the back edges of the eyes (contrast *B. griseocollis, B. fraternus, B. morrisoni, B. rufocinctus*), hindleg basitarsus with the back edge strongly and evenly curved (contrast *B. griseocollis, B. morrisoni, B. fraternus, B. rufocinctus*). Male genitalia with the penis-valve head greatly broadened from the upperside to the underside and flared outward to form (half of) a broad funnel (contrast *B. griseocollis, B. fraternus, B. morrisoni, B. rufocinctus*).

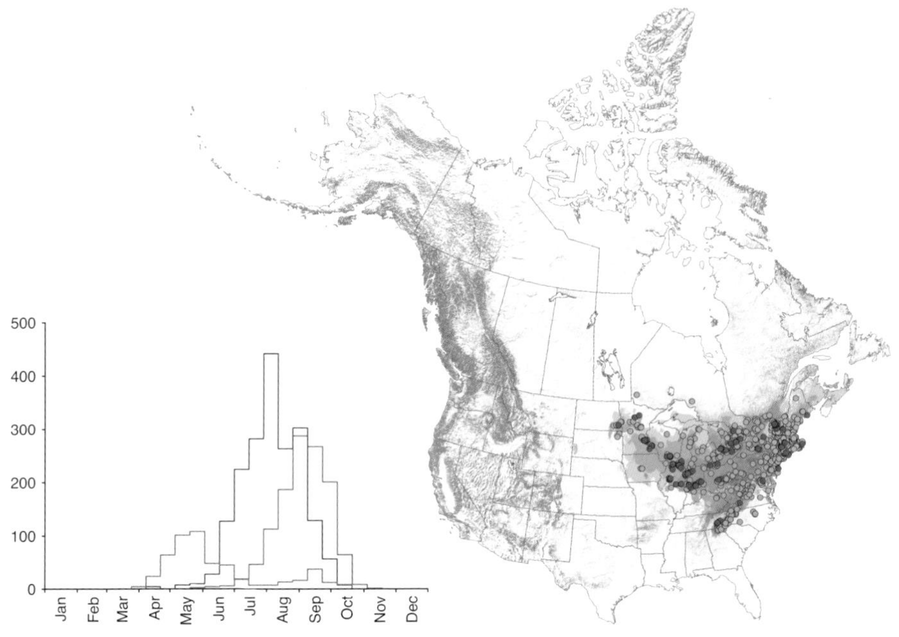

OCCURRENCE

RANGE AND STATUS Northeastern US and adjacent Canada in Eastern Temperate Forest and Boreal Forest regions, south in a narrow band at higher elevations along the Appalachian Mountains, west to the margin of the Great Plains in SD, MN, IA. Formerly common in the Northeast, but after 1996 this species went into rapid and severe decline and is currently very rare. Recent records are mostly from the US Midwest (IL, IN, WI) and southern ON, with very few individuals seen each year since 1997.

HABITAT Close to or within woodland, urban parks and gardens.

EXAMPLE FOOD PLANTS *Aesculus, Agastache, Dalea, Eupatorium, Helianthus, Impatiens, Lonicera, Monarda, Prunus, Solidago, Vaccinium.*

BEHAVIOR Nests underground in deserted mammal burrows (such as those of chipmunks or Eastern cottontail rabbit). A frequent nectar robber of long-corolla flowers. Males patrol circuits in search of mates.

PARASITISM BY OTHER BEES Host to *B. bohemicus* (=*ashtoni*), confirmed breeding record.

BOMBUS GRISEOCOLLIS (DEGEER, 1773)
BROWN-BELTED BUMBLE BEE

LEFT: **Male *Bombus griseocollis*.** LR
ABOVE: **Queen *Bombus griseocollis*.** PW

IDENTIFICATION

Eastern and northern, short-tongued species. Most similar to *B. affinis, B. fraternus, B. morrisoni*, and *B. rufocinctus* (see also *B. perplexus, B. vagans, B. sandersoni, B. bimaculatus*, and *B. impatiens*). The palest color patterns occur in the west.

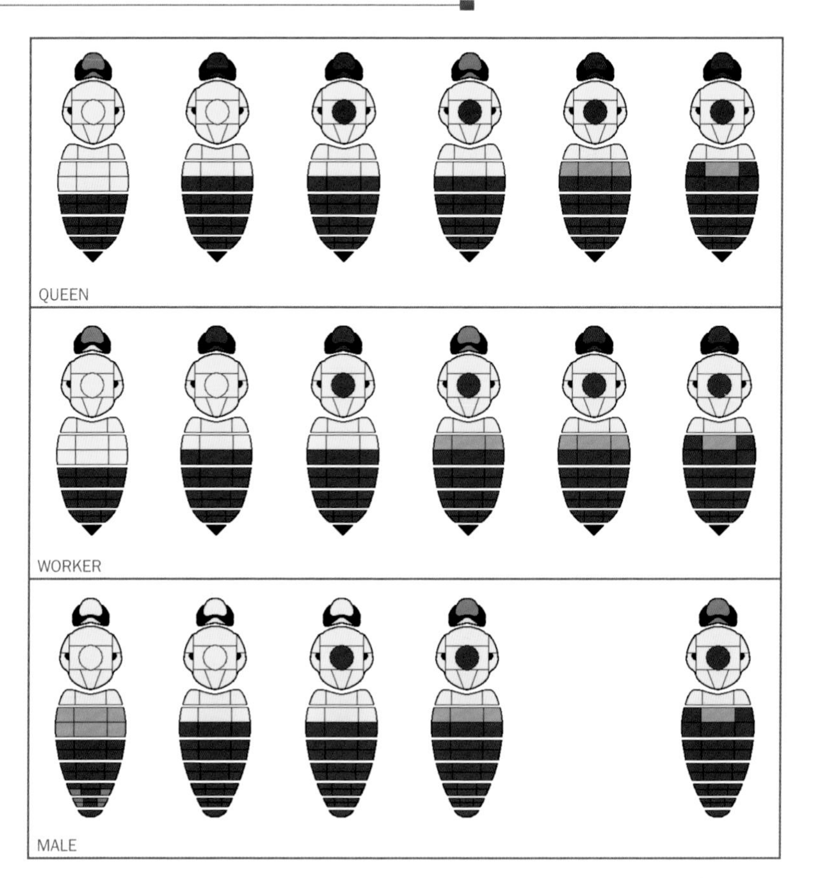
QUEEN
WORKER
MALE

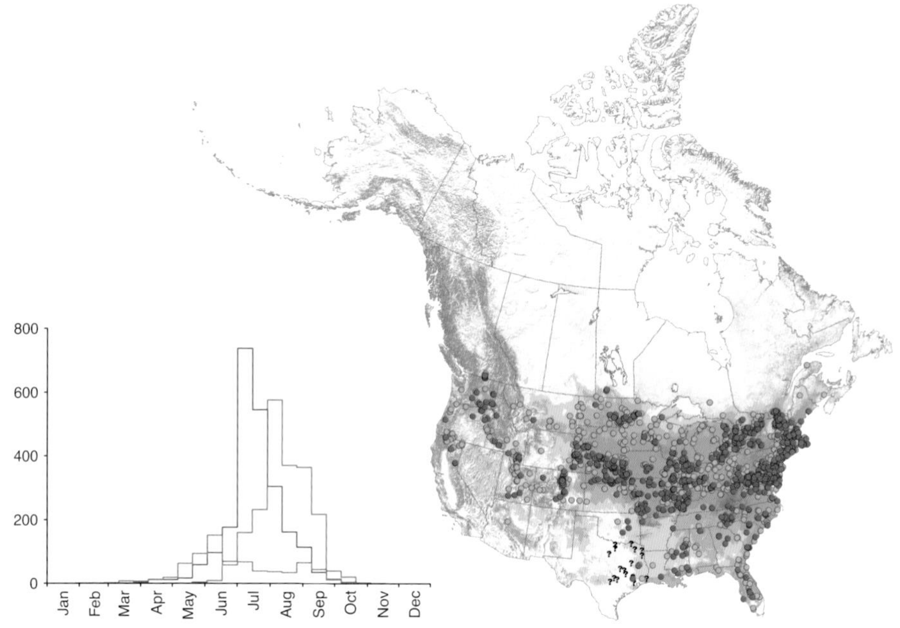
800
600
400
200
0
Jan
Feb
Mar
Apr
May
Jun
Jul
Aug
Sep
Oct
Nov
Dec

HAND CHARACTERS Body size medium: queen 18–23 mm (0.72–0.92 inch), worker 10–16 mm (0.41–0.62 inch). Hair very short and even (contrast *B. affinis*). Head very short with the cheek (oculo-malar area) distinctly shorter than broad (contrast *B. perplexus, B. vagans, B. sandersoni, B. bimaculatus, B. impatiens*), midleg basitarsus with the back far corner rounded, hindleg tibia outer surface flat without long hair but with long fringes at the sides, forming a pollen basket (corbicula). Hair of the face and the upperside of the head black (contrast *B. morrisoni, B. rufocinctus*) or with only a few yellow hairs intermixed, upperside of the thorax between the wings with a black spot often very small and inconspicuous but dense, *sides of the thorax entirely yellow* (contrast *B. morrisoni,* most *B. fraternus*), metasomal T2 often yellow in the queen but brown in the worker (contrast *B. morrisoni, B. fraternus, B. rufocinctus*) usually with black at the back (contrast *B. affinis, B. morrisoni, B. fraternus*). Occasionally worker T2 yellow/brown extending for 0.75× of the length of T2 and forming a W shape similar to *B. bimaculatus.* Rarely T2–5 with orange-red hair at the back, although this may be from damage at the pupal stage. Metasoma rather rectangular and slightly flattened. Male 12–18 mm (0.48–0.70 inch). Eye greatly enlarged, much larger than the eye of any female bumble bee (contrast *B. perplexus, B. vagans, B. sandersoni, B. bimaculatus, B. impatiens*), eyes weakly convergent in the upper part with the upper distance between them more than 0.75× the distance between them below. Antenna long, flagellum nearly 3× longer than the scape. Hair color pattern similar to the queen/ worker, but metasomal T2 usually brownish (contrast *B. fraternus, B. morrisoni, B. rufocinctus*).
MICROSCOPIC CHARACTERS Queen/worker mandible with a very shallow notch in front of the back tooth (contrast *B. affinis*), inner eye margin opposite the ocelli with a few widely spaced small and large pits or punctures occupying a narrow band for less than half the distance from the eye to the neighboring ocellus (contrast *B. rufocinctus*), ocelli large and located in front of a line between the back edges of the eyes (contrast *B. affinis*), hindleg basitarsus with the back edge weakly curved (contrast *B. affinis*). Male ocelli large and located far in front of a line between the back edges of the eyes, the side ocellus less than half an ocellar diameter from the eye. Genitalia with the penis-valve head sickle-shaped (contrast *B. affinis*), the back-curved "sickle" flattened, *no longer than broad as an equilateral triangle* (contrast *B. morrisoni, B. fraternus, B. rufocinctus*), the outer flange of the penis-valve head without a hook (contrast *B. fraternus*), the penis-valve angle on the underside of the shaft and to the side only very weakly marked by a broad curve, gonostylus shorter than broad.

OCCURRENCE

RANGE AND STATUS One of the most widespread and abundant species in the Eastern Temperate Forest and Great Plains regions, also at lower elevations of the Mountain West from northern CA to southern BC.
HABITAT Open farmland and fields, urban parks and gardens, wetlands.
EXAMPLE FOOD PLANTS *Asclepias, Coronilla, Dalea, Echinacea, Lythrum, Melilotus, Monarda, Pontederia, Rudbeckia, Solidago, Trifolium, Verbena, Vicia.*
BEHAVIOR Nests underground or on the surface of the ground. Males perch and chase moving objects in search of mates. Males prefer high perches and have even been found near the top of the Empire State Building (102 stories above ground level).
PARASITISM BY OTHER BEES Unknown.

BOMBUS MORRISONI CRESSON, 1878
MORRISON BUMBLE BEE

Female *Bombus morrisoni.* RHE

Male *Bombus morrisoni.* CD

IDENTIFICATION

Western, short-tongued species. Most similar to *B. griseocollis, B. affinis,* and *B. fraternus* (see also *B. nevadensis, B. perplexus, B. vagans,* and *B. sandersoni*).

HAND CHARACTERS Body size large: queen 22–26 mm (0.87–1.02 inch), worker 12–22 mm (0.47–0.87 inch). Hair very short and even. Head short with the cheek (oculo-malar area) distinctly shorter than broad (contrast *B. nevadensis, B. perplexus, B. vagans, B. sandersoni*), midleg basitarsus with the back far corner rounded, hindleg tibia outer surface flat without long hair but with long fringes at the sides, forming a pollen basket (corbicula). *Hair on the sides of the thorax black* (contrast *B. griseocollis, B. affinis*), upperside of the thorax without a black band between the wings, at most with only a few short inconspicuous black hairs (contrast *B. fraternus*), metasomal T1–2 yellow, T3 with at least some yellow in the middle. Male 15–20 mm (0.58–0.77 inch). Eye greatly enlarged, much larger than the eye of any female bumble bee (contrast *B. perplexus, B. vagans, B. sandersoni*), eyes weakly convergent in the upper part with the upper distance between them more than 0.75× the distance between them below. Antenna long, flagellum nearly 3× longer than the scape (contrast *B. nevadensis*). Hair color pattern similar to the queen/worker, metasomal T3 entirely yellow.

MICROSCOPIC CHARACTERS Queen/worker mandible with the front ridge or keel reaching the far margin (contrast *B. nevadensis*), mandible with a very shallow notch in front of the back tooth (contrast *B. affinis*), clypeus in the middle near the labrum smooth and shining with few pits or punctures (contrast *B. nevadensis*), inner eye margin opposite the ocelli with a few widely spaced small and large punctures occupying a narrow band less than half the distance to the side ocellus (contrast *B. rufocinctus*), ocelli large and located in front of a line between the back edges of the eyes (contrast *B. affinis*), hindleg basitarsus with the back edge weakly curved (contrast *B. affinis*). Male ocelli large and located far in front of a line between the back edges of the eyes, side ocellus less than half an ocellar diameter from the eye. Genitalia with the penis-valve head sickle-shaped (contrast *B. affinis, B. nevadensis*), *the back-curved "sickle" moderately broad, flattened, longer than broad and pointed at the tip* (contrast *B. griseocollis*), the outer flange of the

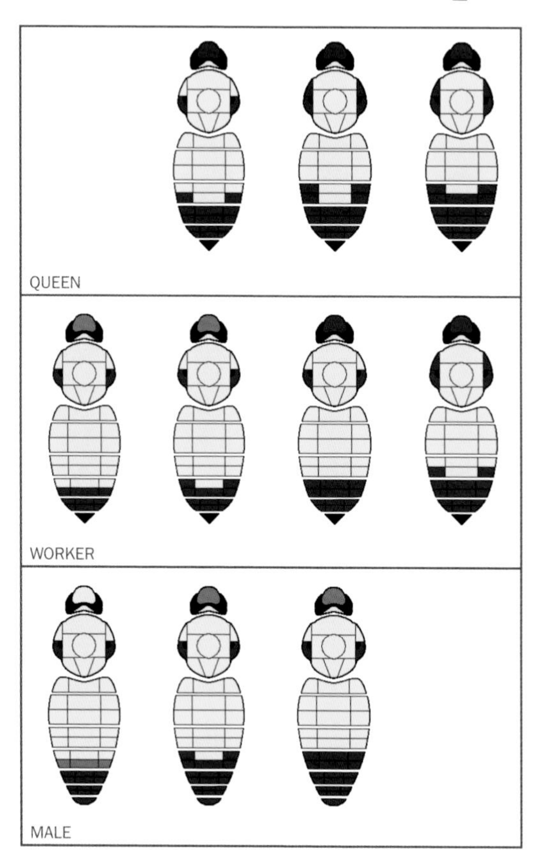

penis-valve head without a hook (contrast *B. fraternus*), the penis valve on the underside of the shaft and to the side straight and without an apparent angle, gonostylus shorter than broad.

OCCURRENCE

RANGE AND STATUS Throughout the US Mountain West from CA east of the Sierra-Cascade Ranges to southern BC, in the Desert West especially in the highlands, and east to NM, TX, and north to western SD. Uncommon, possibly declining in parts of its range.
HABITAT Open dry scrub.
EXAMPLE FOOD PLANTS *Asclepias, Astragalus, Chrysothamnus, Cirsium, Cleome, Ericameria, Helianthus, Melilotus, Senecio.*
BEHAVIOR Nests underground. Males perch and chase moving objects in search of mates.
PARASITISM BY OTHER BEES Unknown.

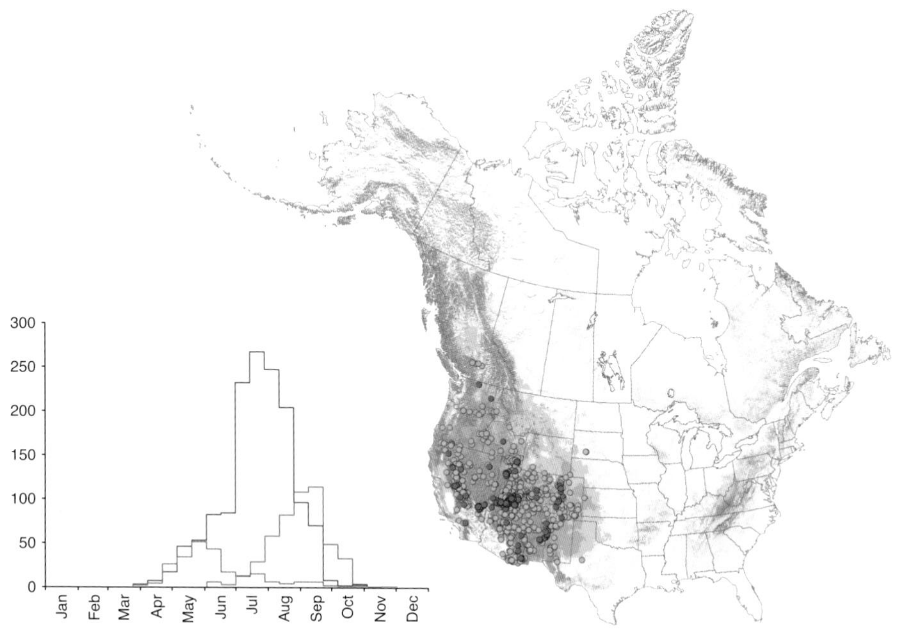

BOMBUS FRATERNUS (SMITH, 1854)
SOUTHERN PLAINS BUMBLE BEE

Worker *Bombus fraternus*. PW

Male *Bombus fraternus*. JCJ

IDENTIFICATION

Southeastern, short-tongued species. Most similar to *B. affinis*, *B. griseocollis*, and *B. rufocinctus* (see also *B. bimaculatus*, *B. vagans*, *B. sandersoni*, and *B. hyperboreus*).

HAND CHARACTERS Body size large (contrast *B. rufocinctus*): queen 25–27 mm (0.97–1.07 inch), worker 15–19 mm (0.56–0.75 inch). Hair very short and even and *hair on metasomal T3 lying completely flat to the body surface* (contrast *B. affinis*, *B. griseocollis*, *B. rufocinctus*). Head very short with the *cheek* (oculo-malar area) *much shorter than broad* (contrast *B. affinis*, *B. bimaculatus*, *B. vagans*, *B. sandersoni*, *B. hyperboreus*), midleg basitarsus with the back far corner rounded, hindleg tibia outer surface flat without long hair but with long fringes at the sides, forming a pollen basket (corbicula). Hair on the *sides of the thorax usually black* (contrast *B. griseocollis*, *B. affinis*), metasomal T1–2 yellow, T3 entirely black. Metasoma nearly rectangular and slightly flattened. Male 22–25 mm (0.85–1.00 inch). Eye greatly enlarged, much larger than the eye of any female bumble bee (contrast *B. bimaculatus*, *B. vagans*, *B. sandersoni*, *B. hyperboreus*), eyes weakly convergent in the upper part with the upper distance between them more than 0.75× the distance between them below. Antenna long, flagellum nearly 3× longer than the scape. Hair color pattern similar to the queen/worker, head and upperside of the thorax between the wings sometimes extensively yellow.

MICROSCOPIC CHARACTERS Queen/worker mandible with a very shallow notch in front of the back tooth (contrast *B. affinis*), inner eye margin opposite the ocelli with a few widely spaced small and large pits or punctures occupying a narrow band for half the distance from the eye to the side ocellus (contrast *B. rufocinctus*), ocelli large and located in front of a line between the back edges of the eyes (contrast *B. affinis*), hindleg basitarsus with the back edge almost straight and parallel to the front edge (contrast *B. affinis*). Male ocelli large and located far in front of a line between the back edges of the eyes, side ocellus less than half an ocellar diameter from the eye, *mandible with the beard on the back margin short and sparse, shorter than the maximum breadth*

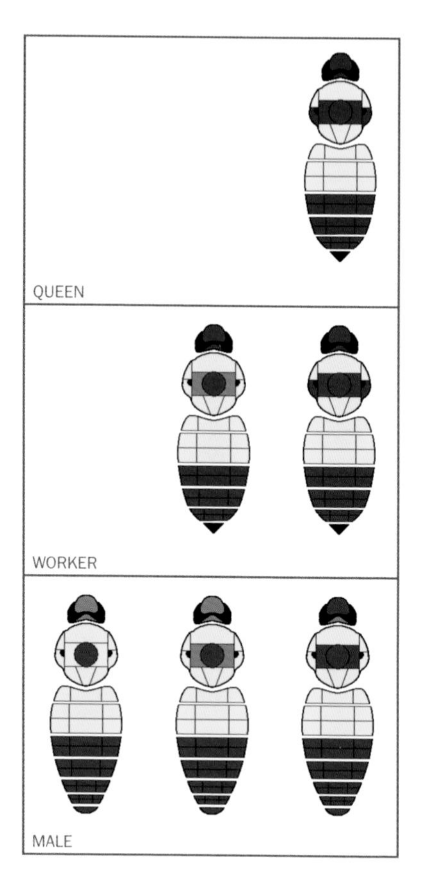

of the mandible (contrast *B. affinis, B. griseocollis, B. rufocinctus*). Genitalia with the penis-valve head sickle-shaped (contrast *B. affinis*), the back-curved "sickle" moderately broad, flattened, but longer than broad and pointed at the tip, the *outer flange of the penis-valve head with a hook at the nearer end of the head* (contrast *B. griseocollis, B. rufocinctus*), the penis-valve angle on the underside of the shaft and to the side almost unmarked and nearly straight, gonostylus as long as broad.

OCCURRENCE

RANGE AND STATUS Eastern Temperate Forest region on the coastal plain of the southeastern US from central FL north to NJ, OH west throughout the US Great Plains. May be declining in parts of its range.

HABITAT Grasslands, urban gardens.

EXAMPLE FOOD PLANTS *Asclepias, Cassia, Dalea, Liatris, Melilotus, Ratibida, Solidago.*

BEHAVIOR Nests underground. Males perch and chase moving objects in search of mates.

PARASITISM BY OTHER BEES Unknown.

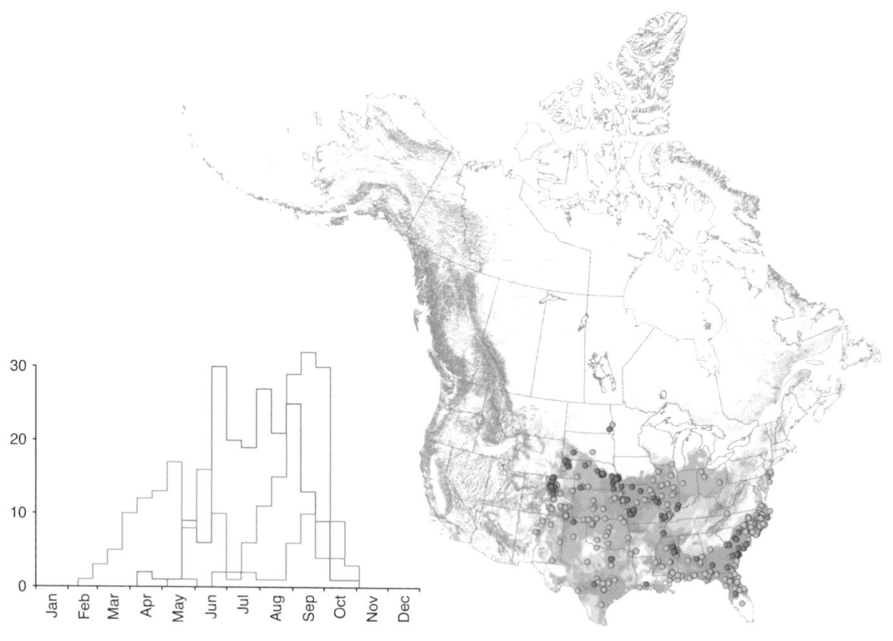

BOMBUS CROTCHII CRESSON, 1878
CROTCH BUMBLE BEE

Queen *Bombus crotchii.* GZ   **Male *Bombus crotchii.*** RH

IDENTIFICATION

Southwestern, short-tongued species. Most similar to *B. griseocollis*, *B. fraternus*, and *B. rufocinctus* (see also *B. frigidus*, some *B. nevadensis*).

HAND CHARACTERS Body size large: queen 22–25 mm (0.87–0.98 inch), worker 12–20 mm (0.47–0.80 inch). Hair very short and even. Head short with the cheek (oculo-malar area) distinctly shorter than broad (contrast *B. frigidus*, *B. nevadensis*), midleg basitarsus with the back far corner rounded, hindleg tibia outer surface flat without long hair but with long fringes at the sides, forming a pollen basket (corbicula). Hair of the upperside of the head at the front black but at the back yellow, *upperside of the thorax at the back* (scutellum), *sides of the thorax, and metasomal T1 predominantly black* (contrast *B. griseocollis*, *B. fraternus*, most *B. rufocinctus*), T2 broadly yellow but often with black in the middle at the front, T5 often red (contrast *B. griseocollis*, *B. fraternus*, *B. rufocinctus*). Metasoma rather rectangular and slightly flattened. Male 14–19 mm (0.56–0.74 inch). Eye greatly enlarged, much larger than the eye of any female bumble bee (contrast *B. frigidus*), eyes weakly convergent in the upper part with the upper distance between them more than 0.75× the distance between them below. Antenna long, flagellum nearly 3× longer than the scape (contrast *B. nevadensis*). Hair color pattern similar to the queen/worker, but the face, upperside of the thorax at the back (scutellum), and metasomal T1 yellow.

MICROSCOPIC CHARACTERS Queen/worker mandible with a very shallow notch in front of the back tooth (contrast *B. affinis*), labrum with the lengthwise middle furrow broader than long (contrast *B. nevadensis*), inner eye margin opposite the ocelli with a few widely spaced small and large pits or punctures occupying a narrow band less than half the distance to the side ocellus (contrast *B. rufocinctus*), ocelli large and located in front of a line between the back edges of the eyes, hindleg basitarsus with the back edge weakly curved (contrast *B. affinis*). Male ocelli large and located

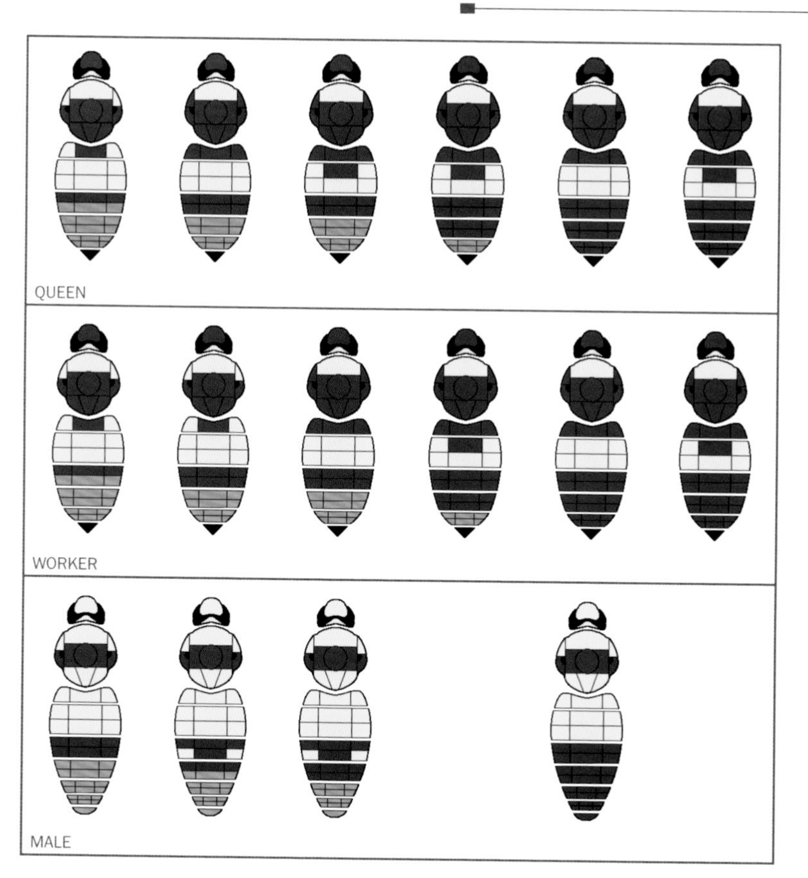
QUEEN
WORKER
MALE

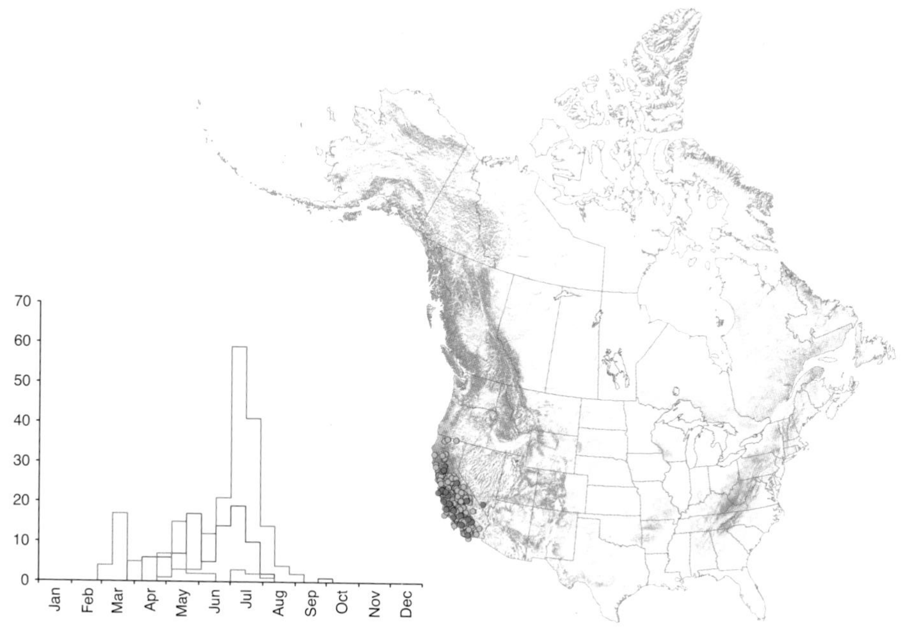
70
60
50
40
30
20
10
0
Jan
Feb
Mar
Apr
May
Jun
Jul
Aug
Sep
Oct
Nov
Dec

far in front of a line between the back edges of the eyes, the side ocellus less than half an ocellar diameter from the eye. Genitalia with the penis-valve head sickle-shaped, the back-curved "sickle" narrow, flattened, much longer than broad and pointed at the tip (contrast *B. griseocollis*), the outer flange of the penis-valve head without a hook near its base (contrast *B. fraternus*), the penis-valve angle on the underside of the shaft and to the side almost unmarked and nearly straight, gonostylus nearly as long as broad with the inner (medial) edge convex, volsella projecting beyond the gonostylus by less than 1× the volsellar breadth (contrast *B. rufocinctus*).

OCCURRENCE

RANGE AND STATUS Primarily CA: Mediterranean, Pacific Coast, Western Desert, Great Valley and adjacent foothills through most of southwestern CA. Moderately common, recorded more frequently in central CA in recent years. Also in Mexico.

HABITAT Open grassland and scrub.

EXAMPLE FOOD PLANTS *Asclepias, Chaenactis, Lupinus, Medicago, Phacelia, Salvia.*

BEHAVIOR Nests underground. Males perch and chase moving objects in search of mates.

PARASITISM BY OTHER BEES Unknown.

BOMBUS RUFOCINCTUS CRESSON, 1863
RED-BELTED BUMBLE BEE

ABOVE: Female *Bombus rufocinctus.* LR
ABOVE RIGHT: Worker *Bombus rufocinctus.* LR
RIGHT: Worker *Bombus rufocinctus.* LR
BELOW: Male *Bombus rufocinctus.* SCO

IDENTIFICATION

Northern and western mountain, short-tongued species. Similar to many other species and often mistaken for them. Colonies usually contain bees with several different color patterns, and there seems to be little geographic pattern in their distribution. Evidence from DNA barcodes confirms that individuals with the different color patterns are conspecific.

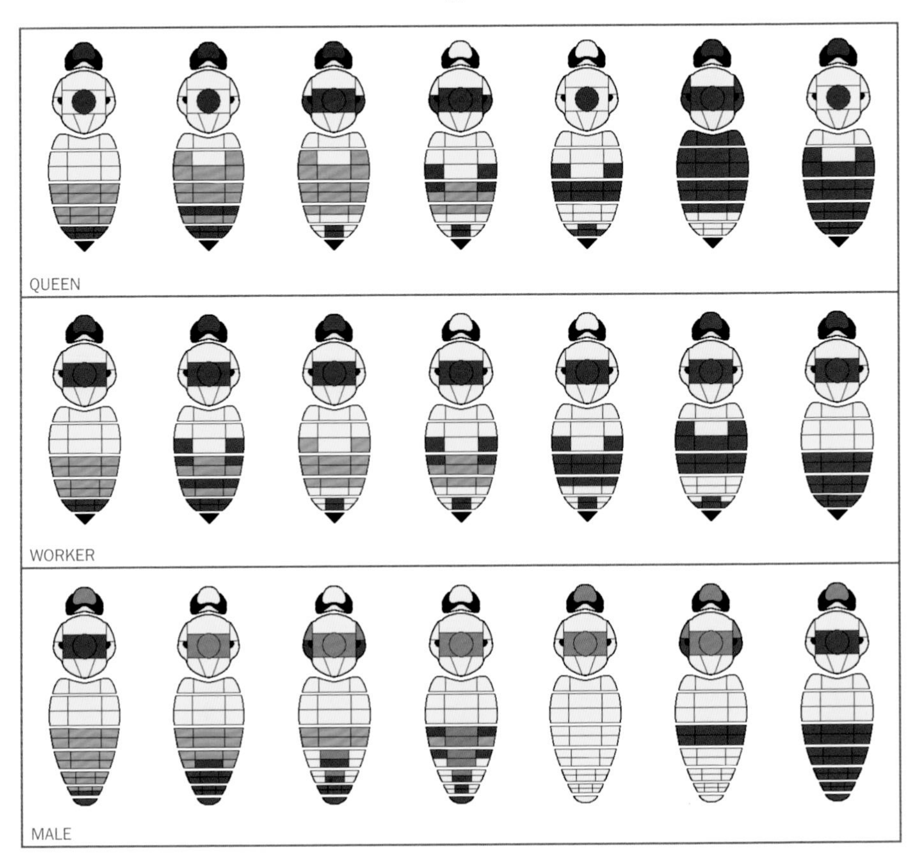
QUEEN
WORKER
MALE

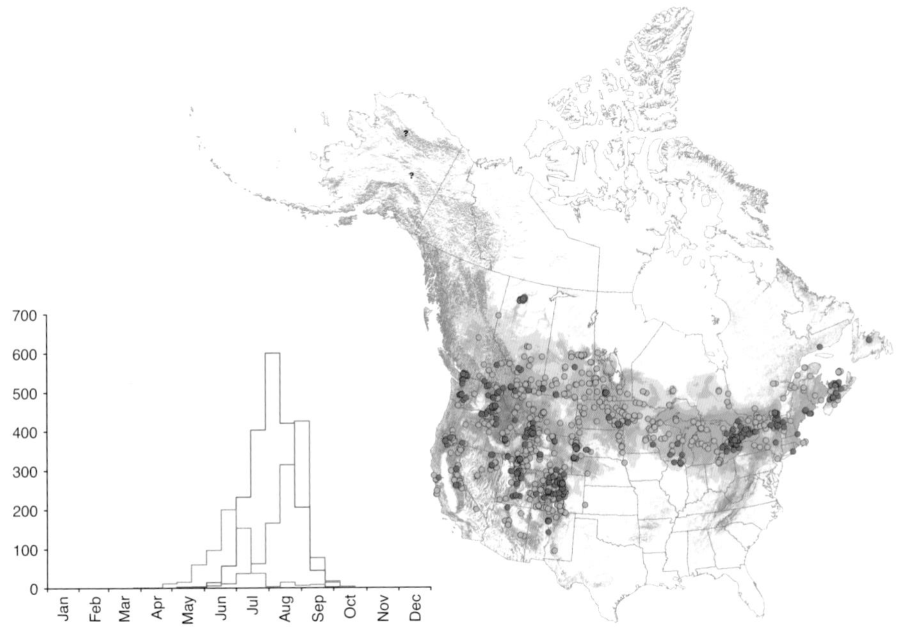
700
600
500
400
300
200
100
0
Jan
Feb
Mar
Apr
May
Jun
Jul
Aug
Sep
Oct
Nov
Dec
?
?

HAND CHARACTERS Body size small: queen 17–18 mm (0.65–0.70 inch), worker 9–13 mm (0.37–0.51 inch). Hair short and even, hair on metasomal T3 erect (contrast *B. fraternus*). Head very short with the *cheek* (oculo-malar area) *distinctly shorter than broad* (contrast most other species, except the much larger and shorter-haired *B. fraternus*), midleg basitarsus with the back far corner rounded, hindleg tibia outer surface flat without long hair but with long fringes at the sides, forming a pollen basket (corbicula). Many other combinations of the color patterns shown here are known, but the hair of metasomal *T2 at the front and near the midline at least usually yellow* (often as a semicircle, only rarely very much reduced; e.g., some individuals with T1–3 black, T4–5 yellow from around San Francisco). Body shape rather globular (contrast *B. fraternus*). Male 11–15 mm (0.42–0.60 inch). Eye slightly enlarged, larger than the eye of any female bumble bee (contrast many similar species), eyes weakly convergent in the upper part with the upper distance between them less than 0.75× the distance between them below. Antenna of medium length, flagellum nearly 3× longer than the scape. Hair color pattern similar to the queen/worker, but the upperside of the thorax black band between the wings often with many yellow hairs intermixed.

MICROSCOPIC CHARACTERS Queen/worker inner eye margin opposite the ocelli with closely spaced small pits or punctures occupying a broad band less than half the distance to the side ocellus (contrast *B. fraternus, B. crotchii, B. griseocollis, B. morrisoni*), ocelli large and located in front of a line between the back edges of the eyes, mandible with a very shallow notch in front of the back tooth (contrast *B. affinis*), hindleg basitarsus with the back edge weakly but evenly arched (contrast *B. affinis*). Male ocelli large and located in front of a line between the back edges of the eyes, about one ocellar diameter from the eye. Genitalia with the penis-valve head sickle-shaped, the back-curved "sickle" narrow, flattened, much longer than broad and pointed at the tip, the outer flange of the penis-valve head without a hook (contrast *B. fraternus*), the penis-valve angle on the underside of the shaft and to the side almost unmarked and nearly straight (contrast many similar species), *gonostylus much shorter than broad and with the inner (medial) edge with two deep concavities, volsella projecting beyond gonostylus by nearly 2× the volsellar breadth* (contrast *B. crotchii*, most other species).

OCCURRENCE

RANGE AND STATUS Canadian Maritimes and Eastern Temperate Forest and Boreal Forest regions, from NL west through the north and central Great Plains, Mountain West, also the Mediterranean and Pacific Coast in CA, with isolated records from AK. Widespread and common throughout its range, although it had disappeared from the San Francisco Bay Area by 2002. Also in Mexico.

HABITAT Close to or within wooded areas, urban parks and gardens.

EXAMPLE FOOD PLANTS *Chicorium, Eupatorium, Fragaria, Grindelia, Helianthus, Melilotus, Solidago, Trifolium, Vicia, Viguiera.*

BEHAVIOR Nests usually on the surface or often aboveground. Males perch and chase moving objects in search of mates.

PARASITISM BY OTHER BEES It is likely that this species is a host to *B. suckleyi*, *B. insularis*, and *B. flavidus* (=*fernaldae*).

BOMBUS FERVIDUS (FABRICIUS, 1798)
YELLOW BUMBLE BEE
(INCLUDING *CALIFORNICUS*)

BELOW: Female *Bombus fervidus*. LR
INSET: Male *Bombus fervidus*. GZ

IDENTIFICATION

Widespread, long-tongued species. Most similar to *B. borealis, B. pensylvanicus, B. appositus, B. distinguendus,* and *B. nevadensis* (see also *B. vosnesenskii, B. caliginosus, B. vandykei, B. perplexus, B. occidentalis, B. franklini, B. insularis,* and *B. flavidus*). Evidence from DNA barcodes supports a close relationship between individuals with the darker color patterns in the west (named *californicus*) and individuals with the lighter color patterns in the east (named *fervidus*). They do not form two separate groups and these individuals show very little barcode divergence, so the two appear to be parts of the same species.

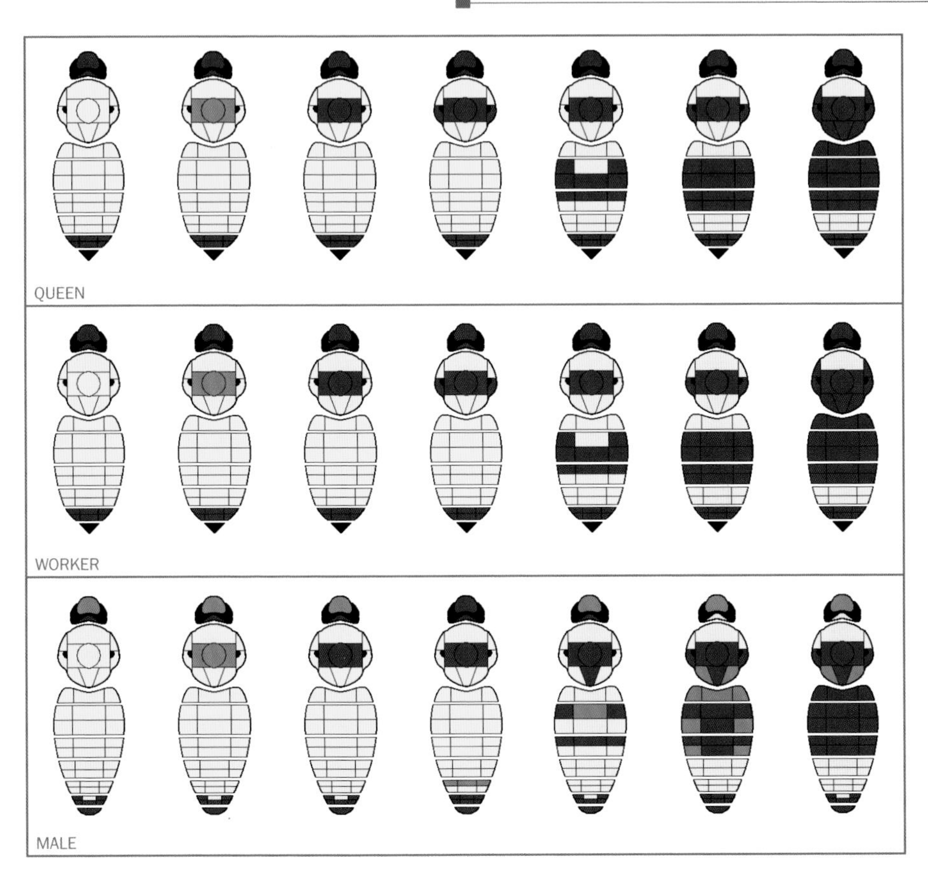
QUEEN
WORKER
MALE

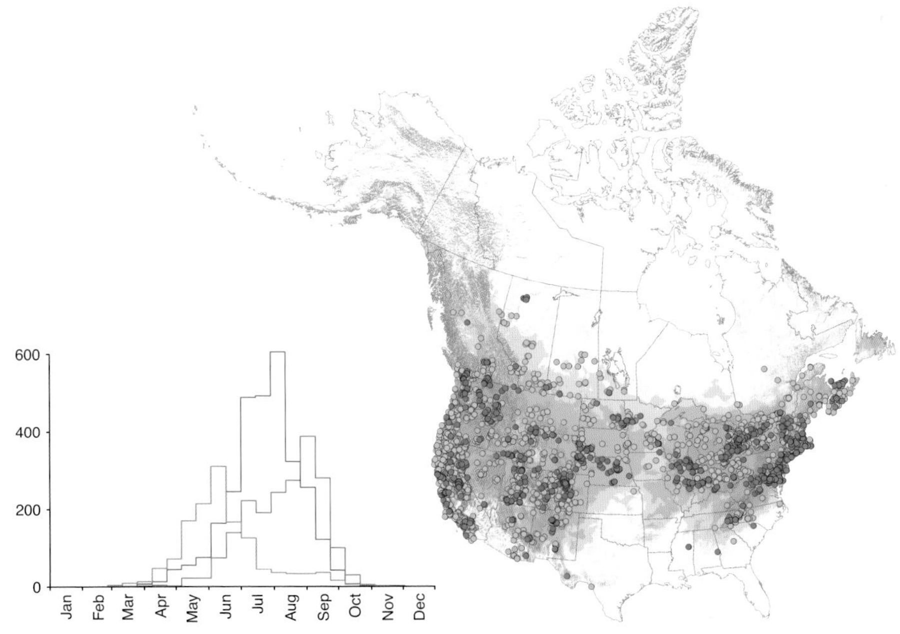
600
400
200
0
Jan
Feb
Mar
Apr
May
Jun
Jul
Aug
Sep
Oct
Nov
Dec

HAND CHARACTERS Body size medium: queen 18–21 mm (0.72–0.84 inch), worker 11–17 mm (0.42–0.67 inch). Hair length medium and even. Head long with the cheek (oculo-malar area) distinctly longer than broad (contrast *B. vosnesenskii, B. caliginosus, B. vandykei, B. perplexus, B. occidentalis, B. franklini*), midleg basitarsus with the back far corner narrowly extended as a spine, hindleg tibia outer surface flat without long hair but with long fringes at the sides, forming a pollen basket (corbicula). Hair of the *face black* (contrast most *B. borealis,* most *B. distinguendus, B. appositus*) or with only a minority of short pale hairs intermixed, upperside of the thorax with the black band between the wings often very narrow (measured from front to back along the midline) and may have many yellow hairs intermixed, or black hairs may be completely lacking, *sides of the thorax yellow* at least in the upper half and often throughout (contrast most *B. borealis, B. appositus, B. distinguendus, B. pensylvanicus*) and either (in the east or northwest) metasomal T1–4 predominantly yellow and T5 black, or (for the darkest color patterns from near the west coast) upperside of the thorax black at the back and T2–3 extensively black. Male 13–16 mm (0.52–0.66 inch). Eye similar in size and shape to the eye of any female bumble bee (contrast *B. nevadensis*). Antenna long, flagellum 4× longer than the scape (contrast *B. nevadensis, B. vosnesenskii, B. caliginosus, B. vandykei, B. perplexus, B. occidentalis, B. franklini*). Hair color pattern similar to the queen/worker, but hair of the *face and the upperside of the head black* with a minority of yellow hairs intermixed (contrast *B. borealis, B. appositus,* most *B. pensylvanicus*), *if* metasomal T1 is yellow, *then* the sides of the thorax are extensively yellow and without black hairs intermixed, and the upperside of the thorax at the back is yellow without black hairs intermixed, T5 usually yellow, T7 usually black or *if* T7 is orange *then* T2–3 are extensively black.
MICROSCOPIC CHARACTERS Queen/worker *clypeus surface rough with many large and small pits or punctures* (contrast *B. borealis, B. appositus, B. distinguendus*), although large punctures are fewer near the midline and near the labrum. Male hindleg tibia outer surface center rough and matte, with many short hairs, hairs of the back fringe shorter than the maximum breadth of the tibia. Genitalia with the *penis-valve head turned outward from the body midline* (contrast *B. borealis, B. appositus, B. distinguendus*) and *about as long as broad, triangular* (contrast *B. pensylvanicus*).

OCCURRENCE

RANGE AND STATUS A widespread species across much of the midlatitudes of the continent, from the Canadian Maritimes and eastern US in Eastern Temperate Forest and Boreal Forest regions, west through the central Great Plains of US and southern Canada to the Mountain West, Pacific Coast and Eastern Desert of CA. Not abundant in the Boreal Forest region and appears to be in slow decline in parts of its range. Also in Mexico.
HABITAT Open grassland, farmland, urban parks and gardens.
EXAMPLE FOOD PLANTS *Astragalus, Cirsium, Helianthus, Lonicera, Lythrum, Monarda, Pedicularis, Penstemon, Trifolium, Vicia.*
BEHAVIOR Nests usually on the surface or often aboveground, in deserted mouse nests, tall grass, or hay stacks, but occasionally underground. One of the more aggressive species, probably as an adaptation to protect the more exposed aboveground nests. Males congregate outside nest entrances in search of mates.
PARASITISM BY OTHER BEES Host to *B. insularis,* confirmed breeding record, likely also host to *B. suckleyi.*

BOMBUS BOREALIS KIRBY, 1837
NORTHERN AMBER BUMBLE BEE

LEFT: **Queen *Bombus borealis*.** PW
ABOVE: **Worker *Bombus borealis*.** LR
BELOW: **Males *Bombus borealis*.** LR

IDENTIFICATION

Northern, long-tongued species. Most similar to *B. fervidus, B. appositus, B. distinguendus,* and *B. pensylvanicus*. The darkest female color patterns are very rare.

HAND CHARACTERS Body size medium: queen 18–22 mm (0.71–0.87 inch), worker 13–15 mm (0.51–0.59 inch). Hair length medium and even. Head long with the cheek (oculo-malar area) distinctly longer than broad, midleg basitarsus with the back far corner narrowly extended in a

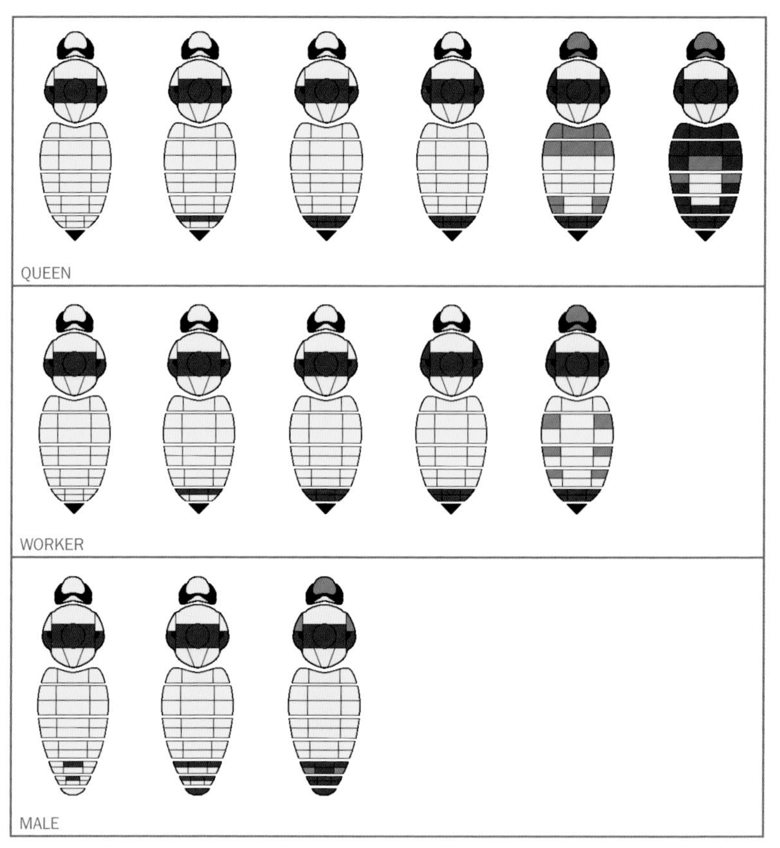

QUEEN
WORKER
MALE

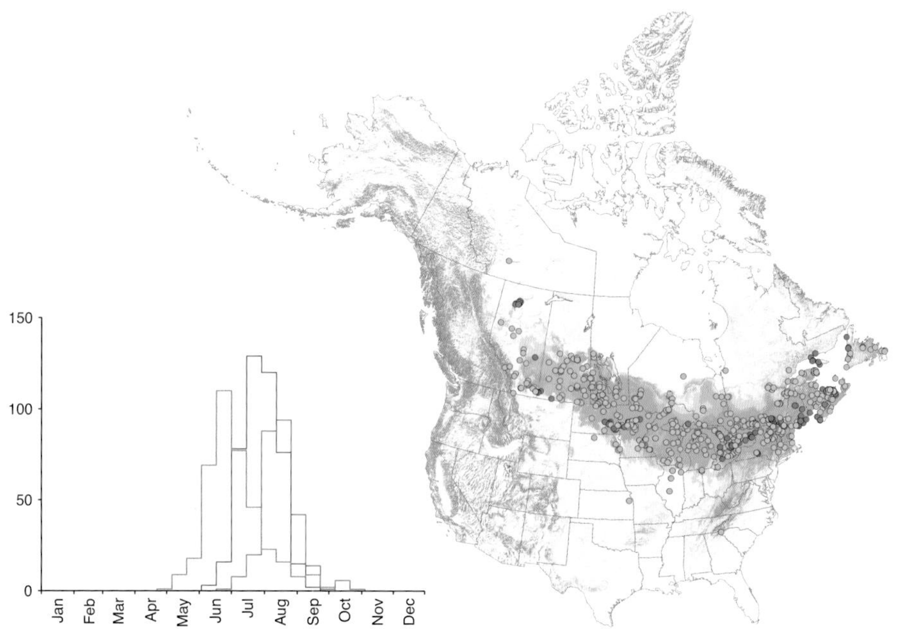

150
100
50
0
Jan
Feb
Mar
Apr
May
Jun
Jul
Aug
Sep
Oct
Nov
Dec

spine, hindleg tibia outer surface flat without long hair but with long fringes at the sides, forming a pollen basket (corbicula). Hair of the face pale (rarely almost black) cream-yellow (contrast *B. fervidus, B. pensylvanicus*), upperside of the head and upperside of the thorax at the front a distinctly darker yellow (contrast *B. appositus, B. distinguendus*), *sides of the thorax predominantly black* with yellow only within the upper half (contrast most eastern *B. fervidus*), metasomal T4 usually an even darker golden yellow (contrast *B. pensylvanicus*), T5 usually with many black hairs (contrast *B. appositus*, most *B. distinguendus*). Male 14–16 mm (0.53–0.62 inch). Eye similar in size and shape to the eye of any female bumble bee. Antenna long, flagellum nearly 4× longer than the scape. Hair color pattern similar to the queen/worker, with the pale hair of the face between the clypeus and the ocelli cream-yellow, hair of the upperside of the head and the upperside of the thorax at the front a distinctly darker sand-yellow, sides of the thorax with many black hairs intermixed throughout, metasomal T1–4 golden yellow, T5 with a black band at least at the front.

MICROSCOPIC CHARACTERS Queen/worker *clypeus surface very smooth and shiny with only few very small pits or punctures near the center* (contrast *B. fervidus, B. pensylvanicus*), metasomal S6 with a lengthwise ridge along the middle (contrast *B. fervidus, B. pensylvanicus*). Male genitalia with the *penis-valve head turned inward as a broad deep spoon shape* (contrast *B. fervidus, B. pensylvanicus*), the penis-valve arm on the underside of the shaft and to the side trident-shaped with the upper tooth strongly extended as a narrow spine and the middle tooth extended as a broad equilateral triangle (contrast *B. distinguendus*).

OCCURRENCE

RANGE AND STATUS Canadian Maritimes and eastern US in Eastern Temperate Forest and Boreal Forest regions, south at high elevations in the Appalachian Mountains, west through the northern Great Plains of ND, MB, AB, and NT.

HABITAT Close to or within wooded areas.

EXAMPLE FOOD PLANTS *Astragalus, Carduus, Cirsium, Melilotus, Rubus, Solidago, Symphytum, Trifolium, Vicia.*

BEHAVIOR Nests underground. Males congregate outside nest entrances in search of mates.

PARASITISM BY OTHER BEES Unknown.

LEFT: **Male *Bombus borealis*.** LR
ABOVE: **Worker *Bombus borealis*.** LR

BOMBUS DISTINGUENDUS MORAWITZ, 1869
NORTHERN YELLOW BUMBLE BEE

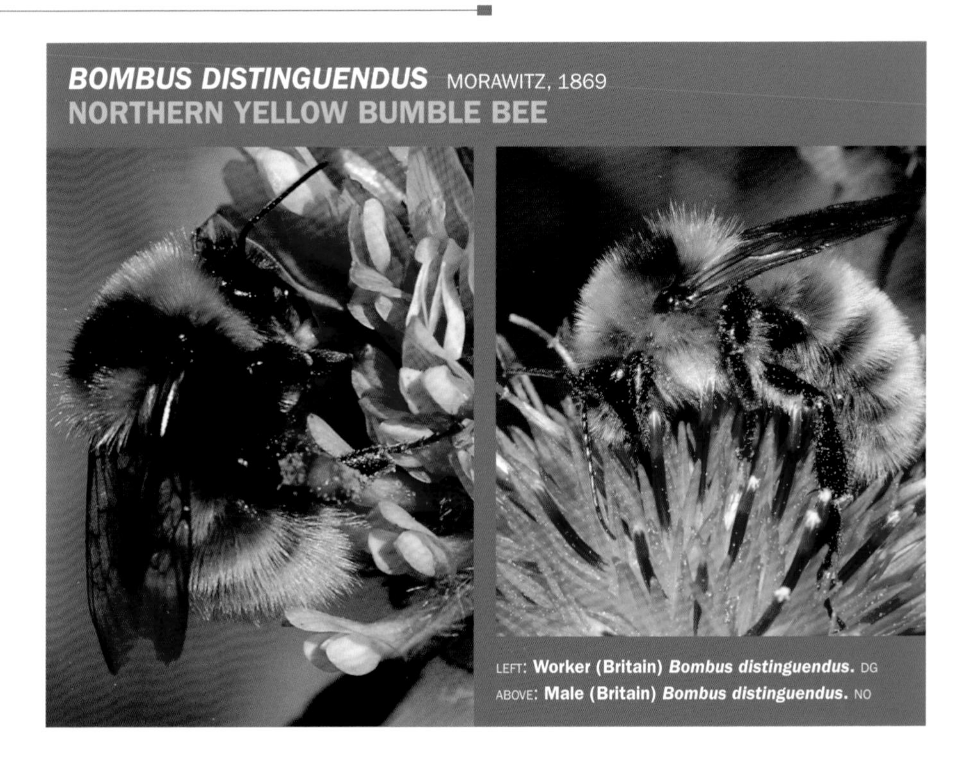

LEFT: **Worker (Britain) *Bombus distinguendus*.** DG
ABOVE: **Male (Britain) *Bombus distinguendus*.** NO

IDENTIFICATION

Far northwestern, long-tongued species. Most similar to *B. fervidus*, *B. appositus*, and *B. borealis*. The first specimen of this species from North America was collected on Attu Island in 2002, but it has since been discovered on mainland AK. This species is widespread but not abundant in northern Europe and Asia. The darkest color patterns are rare and the males are as yet unrecorded from North America.

HAND CHARACTERS Body size medium: queen 19 mm (0.75 inch), worker 12 mm (0.49 inch). Hair length medium and even. Head long with the cheek (oculo-malar area) distinctly longer than broad, midleg basitarsus with the back far corner narrowly extended in a spine, hindleg tibia outer surface flat without long hair but with long fringes at the sides, forming a pollen basket (corbicula). Pale hair of the face usually yellow (contrast *B. fervidus,* most *B. appositus*) and similar to the yellow on metasomal T1 (contrast *B. appositus, B. borealis*) but often with many black hairs extensively intermixed, *sides of the thorax predominantly black* with yellow only within the upper half (contrast most *B. fervidus*), T5 extensively yellow and usually without many black hairs (contrast *B. fervidus,* most *B. borealis*). Male not seen from North America yet. Eye similar in size and shape to the eye of any female bumble bee. Antenna long, flagellum nearly 4× longer than the scape. Hair color pattern similar to the queen/worker.

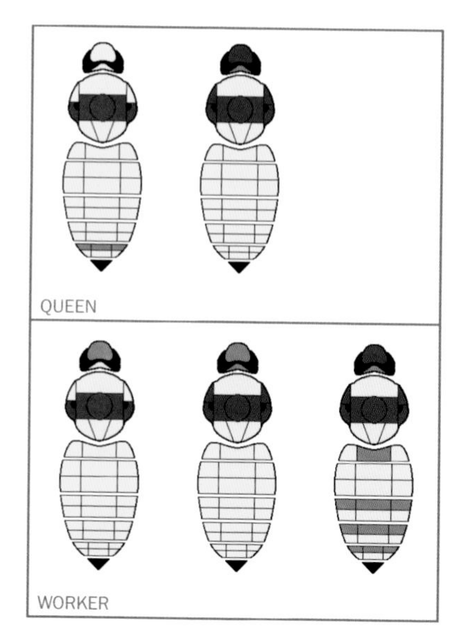

MICROSCOPIC CHARACTERS Queen/worker *clypeus surface very smooth and shiny with only few very small pits or punctures near the center* (contrast *B. fervidus*), metasomal S6 with a ridge along the midline (contrast *B. fervidus*). Male genitalia with the *penis-valve head turned inward as a broad deep spoon shape* (contrast *B. fervidus*), the penis-valve arm on the underside of the shaft and to the side trident-shaped but with *the upper tooth very small and the middle tooth broadly rounded and only weakly marked* (contrast *B. appositus, B. borealis*).

OCCURRENCE

RANGE AND STATUS Tundra/Taiga of AK Aleutian Island archipelago and scattered inland. Probably more broadly distributed in AK than records imply. Also in Europe and Asia.
HABITAT Tundra/Taiga.
EXAMPLE FOOD PLANTS *Aconitum, Geranium, Hedysarum, Lupinus.*
BEHAVIOR Nests underground. Males congregate outside nest entrances in search of mates.
PARASITISM BY OTHER BEES Unknown.

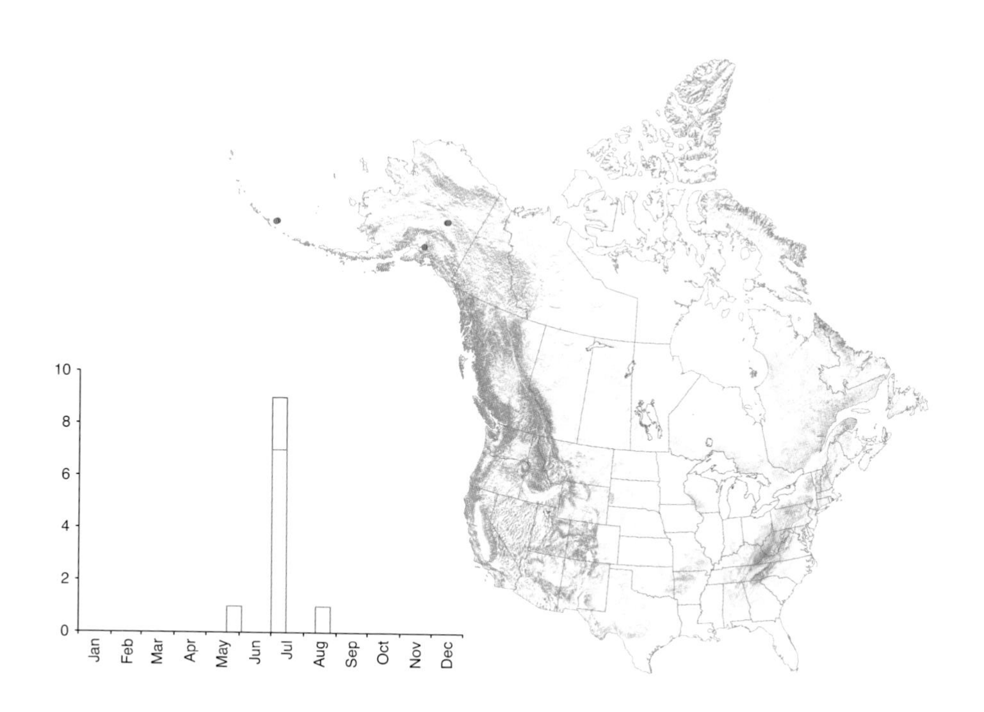

BOMBUS APPOSITUS CRESSON, 1878
WHITE-SHOULDERED BUMBLE BEE

ABOVE: **Female *Bombus appositus*.** DI
LEFT: **Male *Bombus appositus*.** LRE
BELOW: **Male *Bombus appositus*.** LRE

IDENTIFICATION

Western, long-tongued species. Most similar to *B. fervidus, B. borealis,* and *B. distinguendus*. The color pattern in which all the pale hair is yellow is restricted to the region around Corvallis and Ashland, OR. So far no DNA barcodes have been available to confirm that it is conspecific with typical *appositus*, an interpretation based here on similar morphology.

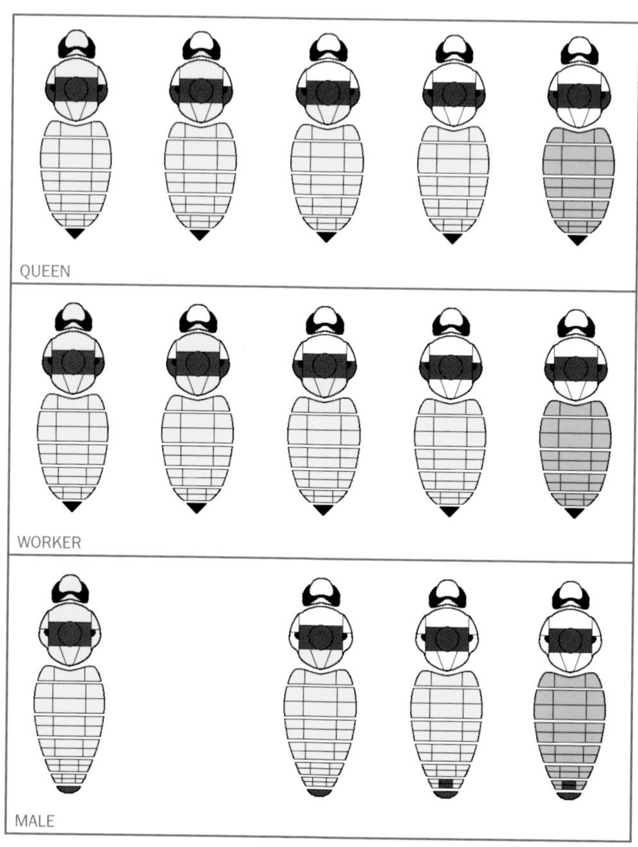
QUEEN
WORKER
MALE

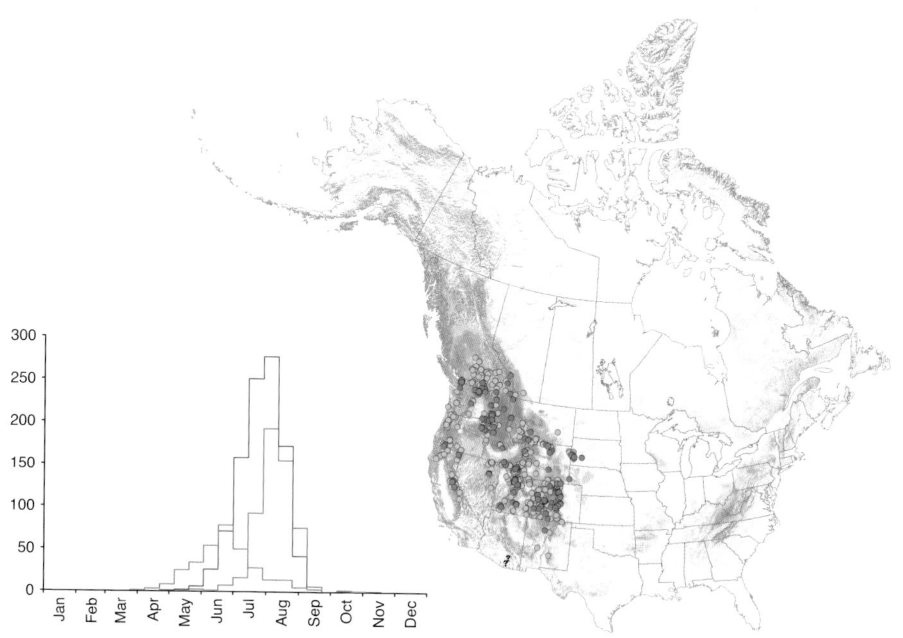
300
250
200
150
100
50
0
Jan
Feb
Mar
Apr
May
Jun
Jul
Aug
Sep
Oct
Nov
Dec

HAND CHARACTERS Body size large: queen 20–24 mm (0.78–0.95 inch), worker 11–18 mm (0.45–0.71 inch). Hair length medium and even. Head long with the cheek (oculo-malar area) distinctly longer than broad, midleg basitarsus with the back far corner narrowly extended in a spine, hindleg tibia outer surface flat without long hair but with long fringes at the sides, forming a pollen basket (corbicula). Hair of the face pale and usually gray-white (contrast *B. fervidus, B. borealis, B. distinguendus*) and usually paler than the hair on metasomal T1 (contrast *B. fervidus, B. borealis, B. distinguendus*) but instead may be yellow, *sides of the thorax predominantly black* with pale hair only in the upper half (contrast most *B. fervidus*), T5 entirely orange-brown or straw yellow (contrast *B. fervidus,* most *B. borealis*). Male 10–16 mm (0.41–0.62 inch). Eye similar in size and shape to the eye of any female bumble bee. Antenna long, flagellum nearly 4× longer than the scape. Hair color pattern similar to the queen/worker, but sides of the thorax with only a few black hairs intermixed toward the back, metasomal T1–5 orange-brown, T5 without black, T6 pale at the sides (contrast *B. fervidus,* most *B. borealis*).
MICROSCOPIC CHARACTERS Queen/worker *clypeus surface very smooth and shiny with only few very small pits or punctures near the center* (contrast *B. fervidus*), metasomal S6 with a ridge along the midline (contrast *B. fervidus*). Male genitalia with the *penis-valve head turned inward as a broad deep spoon shape* (contrast *B. fervidus*), the penis-valve arm on the underside of the shaft and to the side trident-shaped and with the upper tooth strongly extended as a narrow spine and the middle tooth extended as a broad equilateral triangle (contrast *B. distinguendus*).

OCCURRENCE

RANGE AND STATUS Mountain West from NM to southern AB and BC, highlands within the Western Desert, and Sierra Nevada of CA.
HABITAT Open meadows and granitic soil slopes.
EXAMPLE FOOD PLANTS *Agastache, Cirsium, Delphinium, Gentiana, Orthocarpus, Oxytropis, Penstemon, Trifolium.*
BEHAVIOR Nests underground or slightly more often on the surface. Males congregate outside nest entrances in search of mates.
PARASITISM BY OTHER BEES Host to *B. insularis*, confirmed breeding record. It is likely that this species is a host to *B. suckleyi* and *B. flavidus* (=*fernaldae*).

BOMBUS PENSYLVANICUS (DEGEER, 1773)
AMERICAN BUMBLE BEE (INCLUDING *SONORUS*)

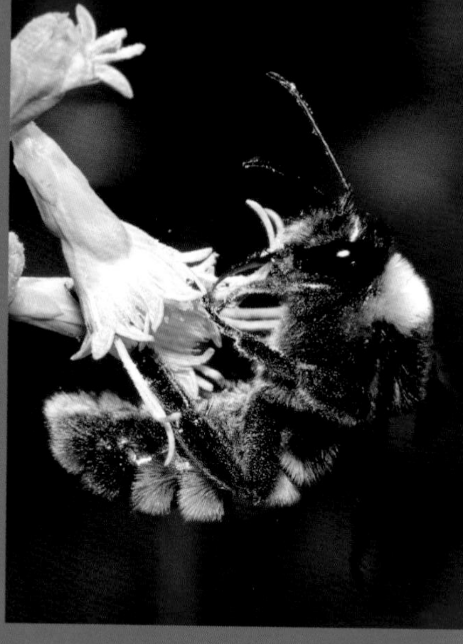

ABOVE LEFT: **Worker *Bombus pensylvanicus*.** PW
ABOVE RIGHT: **Male *Bombus pensylvanicus*.** TL
RIGHT: **Female *Bombus pensylvanicus*.** RH
FAR RIGHT: **Male *Bombus pensylvanicus*.** RH

IDENTIFICATION

Widespread, long-tongued species. Most similar to *B. fervidus, B. borealis, B. appositus, B. distinguendus, B. auricomus*, and *B. nevadensis* (see also *B. terricola, B. occidentalis*, and *B. morrisoni*). Evidence from DNA barcodes supports the individuals with lighter color pattern in the southwest (named *sonorus,* from California) forming a group separate from those with the darker color pattern in the east (named *pensylvanicus*). However, DNA barcodes are available from only a very few sites and do not represent the areas of northern Mexico where studies have reported frequent intermediate color patterns, so that the two groups within the United States appear to be the opposite ends of a cline joined by intermediate individuals within Mexico. Some regard these groups as subspecies.

HAND CHARACTERS Body size large: queen 22–26 mm (0.86–1.01 inch), worker 13–19 mm (0.52–0.76 inch). Hair short and even. Head long with the cheek (oculo-malar area) just longer than broad (contrast *B. terricola, B. occidentalis, B. morrisoni*), midleg basitarsus with the back far corner narrowly extended as a spine, hindleg tibia outer surface flat without long hair but with

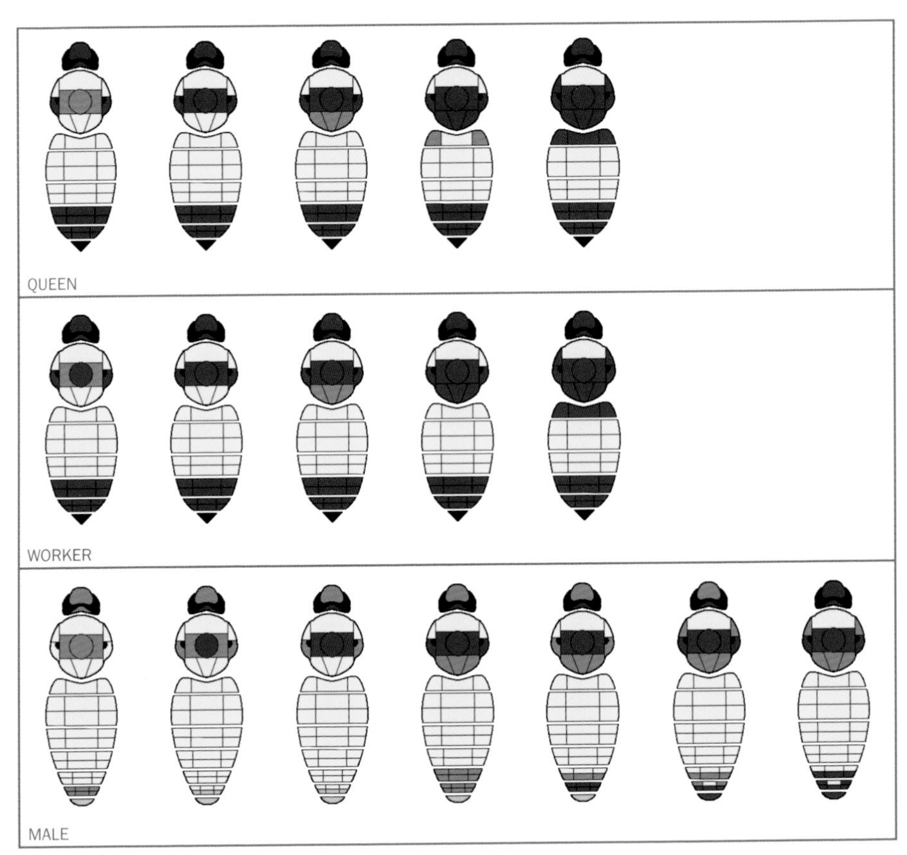
QUEEN
WORKER
MALE

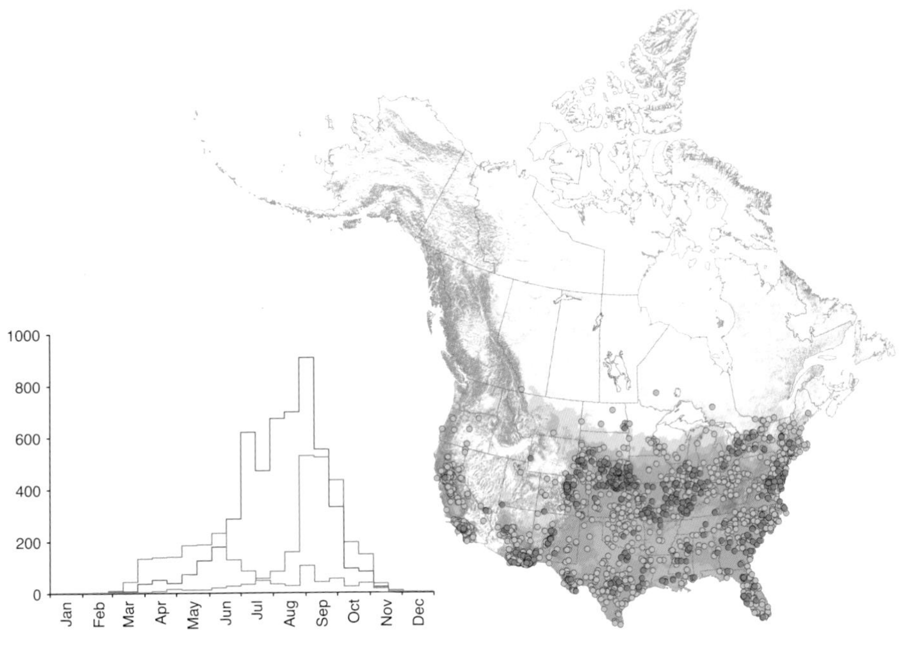
1000
800
600
400
200
0
Jan
Feb
Mar
Apr
May
Jun
Jul
Aug
Sep
Oct
Nov
Dec

long fringes at the sides, forming a pollen basket (corbicula). Hair of the *face and upperside of the head always black* (contrast *B. borealis, B. appositus, B. distinguendus, B. auricomus, B. nevadensis*), sides of the thorax black (contrast eastern *B. fervidus*), metasomal *T1 with yellow hairs more dominant near the midline* (contrast *B. auricomus, B. nevadensis*), T4 black (contrast *B. fervidus*). Male 15–21 mm (0.58–0.84 inch). Eye similar in size and shape to the eye of any female bumble bee (contrast *B. auricomus, B. nevadensis, B. morrisoni*). Antenna long, flagellum 4× longer than the scape (contrast *B. terricola, B. occidentalis*). Hair color pattern similar to the queen/worker, but metasomal T7 often orange (contrast *B. fervidus*), or *if* T7 is black *then* T2–3 are entirely yellow, face with a minority of yellow hairs intermixed, upperside of the head predominantly black, sides of the thorax with black hairs intermixed, upperside of the thorax with the black band between the wings sometimes with yellow hairs intermixed, upperside of the thorax with the back pale band with usually at least a few black hairs intermixed.

MICROSCOPIC CHARACTERS Queen/worker mandible with the front ridge or keel reaching the far margin (contrast *B. auricomus, B. nevadensis*), *clypeus surface rough* with many dense very large pits or punctures, although these are slightly fewer on the midline and near the labrum (contrast *B. borealis, B. appositus, B. distinguendus*), hindleg basitarsus with the back near lobe (adjacent to the tibia) long and pointed and longer than broad (contrast *B. auricomus, B. nevadensis*). Male hindleg tibia outer surface center rough and matte, with many short hairs, hairs of the back fringe shorter than the maximum breadth of the tibia. Genitalia with the *penis-valve head turned outward from the body midline* (contrast *B. borealis, B. appositus, B. distinguendus, B. auricomus, B. nevadensis*) and *more than twice as long as broad, banana-shaped* (contrast *B. fervidus*).

OCCURRENCE

RANGE AND STATUS Widespread in the Eastern Temperate Forest and Great Plains regions throughout the eastern and central US and extreme southern Canada, absent from much of the Mountain West, but found in the Desert West and adjacent areas of CA and OR. Formerly one of the most widespread species in the southern US, but now in decline in the northern parts of its range. Also in Mexico.

HABITAT Open farmland and fields.

EXAMPLE FOOD PLANTS *Astragalus, Cirsium, Cornus, Dalea, Echinacea, Helianthus, Kallstroemia, Liatris, Mentzelia, Silphium, Solanum, Trifolium, Vicia.*

BEHAVIOR Nests mostly on the surface of the ground, among long grass, but occasionally underground. This is one of the more aggressive bumble bee species, probably as an adaptation to protect the more exposed aboveground nests. Males congregate outside nest entrances in search of mates.

PARASITISM BY OTHER BEES Host to *B. variabilis*, confirmed breeding record.

BOMBUS AURICOMUS (ROBERTSON, 1903)
BLACK AND GOLD BUMBLE BEE

LEFT: **Worker *Bombus auricomus*.** PW
ABOVE: **Male *Bombus auricomus*.** TB

IDENTIFICATION

Eastern, long-tongued species. Most similar to *B. nevadensis* and *B. pensylvanicus* (see also *B. terricola* and *B. occidentalis*). This species is replaced by *B. nevadensis* in the west.

HAND CHARACTERS Body size large: queen 22–26 mm (0.86–1.03 inch), worker 16–19 mm (0.64–0.73 inch). Hair very short and even. Head long with the cheek (oculo-malar area) longer than broad (contrast *B. terricola, B. occidentalis*), midleg basitarsus with the back far corner acutely pointed but not narrowly produced, hindleg tibia outer surface flat without long hair but with long fringes at the sides, forming a pollen basket (corbicula). Hair on the face black, occasionally with a few yellow hairs intermixed, *upperside of the head usually yellow* (contrast *B. pensylvanicus*) sometimes with black hairs intermixed but occasionally predominantly black, upperside of the thorax between the wings with a strong black band (contrast *B. nevadensis*), metasomal *T1 with yellow hairs more dominant at the sides* (contrast *B. pensylvanicus*). Body robust and rectangular. Male 13–21 mm (0.52–0.83 inch). Eye greatly enlarged, much larger than the eye of any female bumble bee (contrast *B. pensylvanicus, B. terricola, B. occidentalis*), eyes strongly convergent in the upper part with the upper distance between them half the distance between them below. Antenna short, flagellum 2× longer than the scape (contrast *B. pensylvanicus*). Hair color pattern similar to the queen/worker, but upperside of the head predominantly yellow, upperside of the thorax between the wings with the black band often with yellow intermixed, metasomal *T6–7 black* (contrast *B. nevadensis*).

MICROSCOPIC CHARACTERS Queen/worker *mandible with the front ridge or keel not reaching the far margin* (contrast *B. pensylvanicus, B. terricola*), clypeus very evenly covered with a dense mixture of small and large deep pits or punctures, ocelli large and located in front of a line between the back margins of the eyes (contrast *B. pensylvanicus*), inner eye edge opposite the ocelli with a band of large punctures with the spaces between them rough and dull (contrast *B. nevadensis*), hindleg basitarsus with the back near lobe (adjacent to the tibia) broad and blunt and shorter

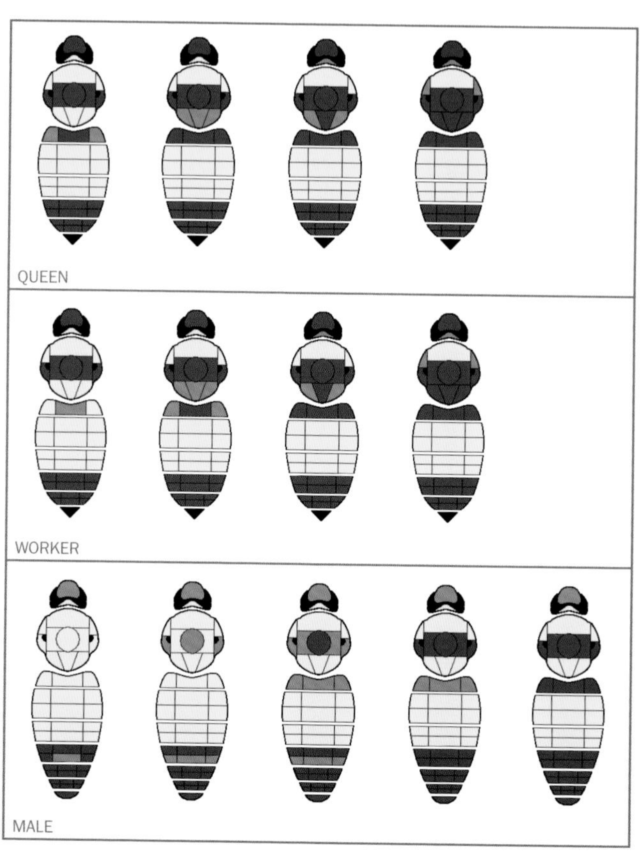
QUEEN
WORKER
MALE

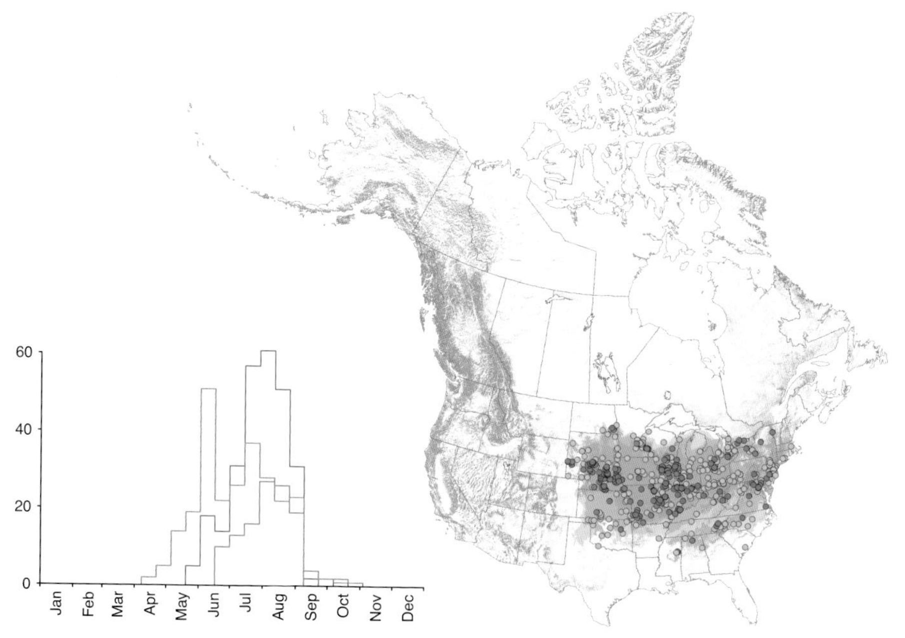
60
40
20
0
Jan
Feb
Mar
Apr
May
Jun
Jul
Aug
Sep
Oct
Nov
Dec

than broad (contrast *B. pensylvanicus, B. terricola*). Male ocelli large and located far in front of a line between the back edges of the eyes, the side ocellus less than half an ocellar diameter from the eye, *the inner eye margin opposite the ocelli with a band of large punctures with the spaces between them rough and dull* (contrast *B. nevadensis*). Genitalia with the *penis-valve head straight and about 5× as long as broad* (contrast *B. pensylvanicus, B. terricola, B. occidentalis*).

OCCURRENCE

RANGE AND STATUS Throughout most of the Eastern Temperate Forest region but scarce in the coastal southeastern US, west to the eastern Great Plains.

HABITAT Open farmland and fields.

EXAMPLE FOOD PLANTS *Carduus, Cirsium, Dalea, Delphinium, Dipsacus, Echinacea, Hypericum, Monarda, Penstemon, Trifolium, Vicia.*

BEHAVIOR Nests are rather small and on the ground surface. Males perch and chase moving objects in search of mates.

PARASITISM BY OTHER BEES Unknown.

BOMBUS NEVADENSIS CRESSON, 1874
NEVADA BUMBLE BEE

LEFT: Female *Bombus nevadensis.* AF
ABOVE: Male *Bombus nevadensis.* WE

IDENTIFICATION

Western, long-tongued species. Most similar to *B. auricomus, B. pensylvanicus,* and *B. fervidus* (see also *B. morrisoni, B. griseocollis, B. terricola, B. occidentalis,* and *B. crotchii*). The darkest female color pattern with a black band between the wings and a black tail (similar to *B. auricomus*) is confined to Vancouver Island, BC. The females with more extensive orange on the metasoma (resembling some *B. crotchii* and some more northern *B. occidentalis* males) are confined to Humboldt County, and San Miguel Island, CA. This species is replaced by *B. auricomus* in eastern North America.

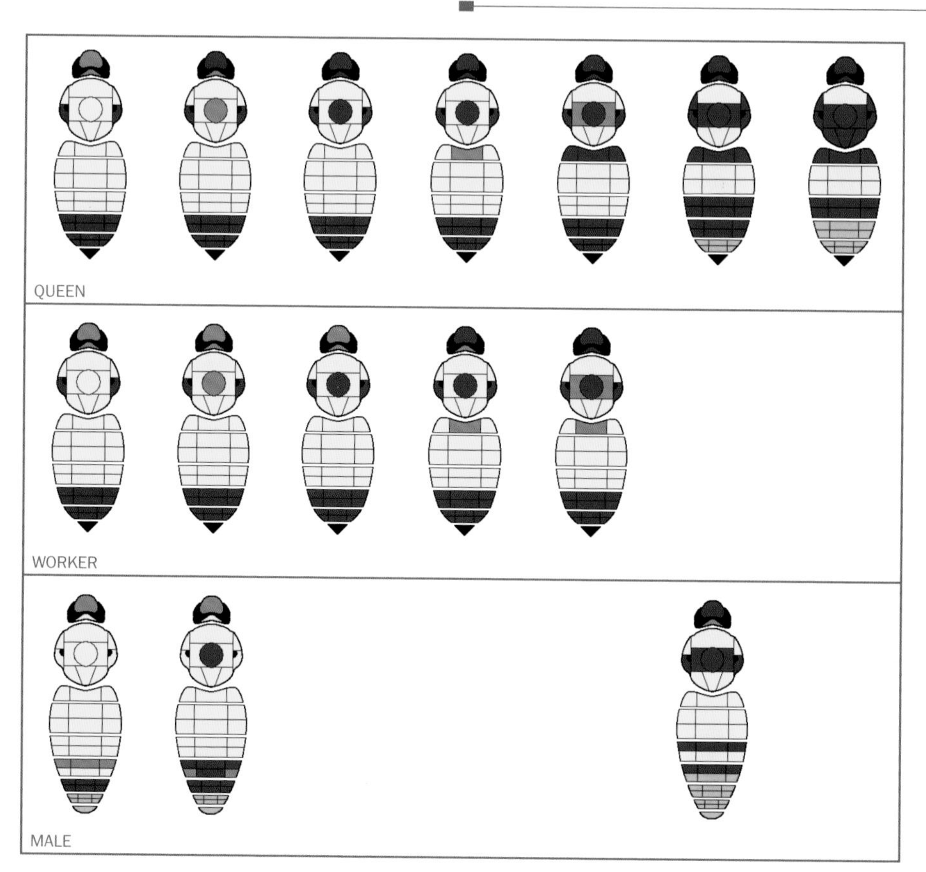
QUEEN
WORKER
MALE

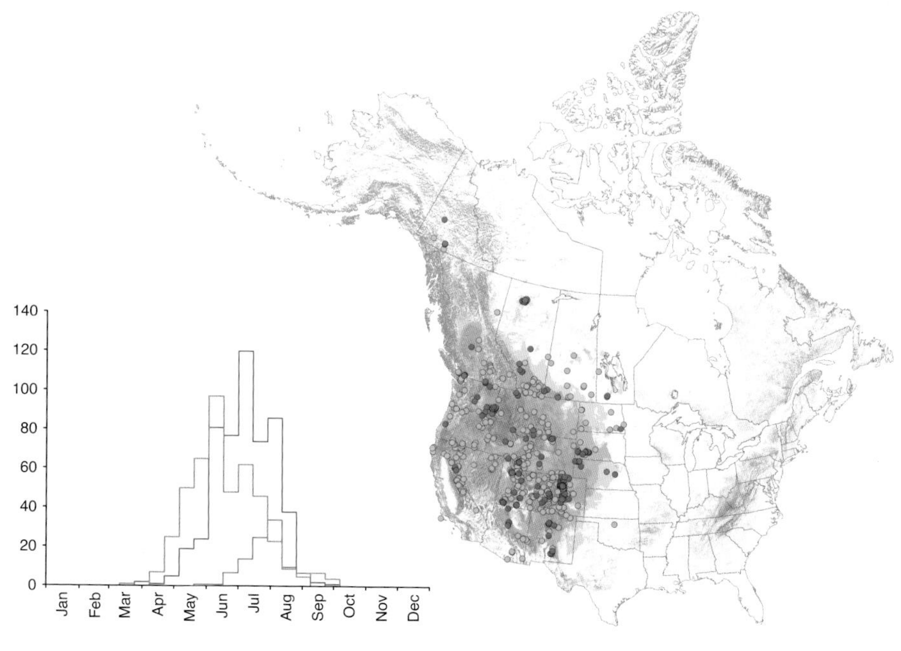
140
120
100
80
60
40
20
0
Jan
Feb
Mar
Apr
May
Jun
Jul
Aug
Sep
Oct
Nov
Dec

HAND CHARACTERS Body size large: queen 24–25 mm (0.93–1.00 inch), worker 15–21 mm (0.58–0.81 inch). Hair very short and even. Head long, with the cheek (oculo-malar area) longer than broad (contrast *B. morrisoni, B. terricola, B. occidentalis, B. crotchii*), midleg basitarsus with the back far corner acutely pointed but not narrowly produced, hindleg tibia outer surface flat without long hair but with long fringes at the sides, forming a pollen basket (corbicula). Hair on the face and the upperside of the head black sometimes with a minority of yellow hairs intermixed, *upperside of the thorax between the wings with yellow usually strongly intermixed in the black spot or band* (contrast *B. auricomus*, most *B. pensylvanicus*, most *B. fervidus*), sides of the thorax predominantly black (contrast most *B. fervidus*), metasomal T1 with yellow hairs more dominant at the sides (contrast *B. pensylvanicus*). Body robust and rectangular. Male 16–19 mm (0.61–0.75 inch). Eye greatly enlarged, much larger than the eye of any female bumble bee (contrast *B. pensylvanicus, B. fervidus, B. terricola, B. occidentalis*), eyes strongly convergent in the upper part with the upper distance between them half the distance between them below. Antenna short, flagellum 2× longer than the scape (contrast *B. pensylvanicus, B. fervidus, B. morrisoni*). Hair color pattern similar to the queen/worker, but metasomal *T6–7 extensively orange* (contrast *B. auricomus, B. fervidus*).
MICROSCOPIC CHARACTERS Queen/worker *mandible with the front ridge or keel not reaching the far margin* (contrast *B. pensylvanicus, B. fervidus*), labrum with the lengthwise midline furrow much narrower than long (contrast *B. crotchii*), clypeus very evenly covered with a dense mixture of small and large deep pits or punctures (contrast *B. morrisoni*), ocelli large and located in front of a line between the back edges of the eyes (contrast *B. pensylvanicus, B. fervidus*), inner eye margin opposite the ocelli with a band of large punctures with the spaces between them smooth and shining (contrast *B. auricomus*), hindleg basitarsus with the back near lobe (adjacent to the tibia) broad and blunt and shorter than broad (contrast *B. pensylvanicus, B. fervidus*). Male ocelli large and located far in front of a line between the back edges of the eyes, the side ocellus less than half an ocellar diameter from the eye, the *inner eye margin opposite the ocelli with a band of large punctures with the spaces between them smooth and shining* (contrast *B. auricomus*). Genitalia with the *penis-valve head straight and about 5× as long as broad* (contrast *B. pensylvanicus, B. fervidus*).

OCCURRENCE

RANGE AND STATUS Mountain West from southern AZ, NM north to BC, YT, highlands of the Desert West, southern CA Transverse Ranges, Sierra-Cascade Ranges and sparingly in the Coast Ranges, east to the western Great Plains. From sea level up to 2,200 m. Moderately common. Has been recorded as present in Mexico.
HABITAT Open grassy prairies and meadows.
EXAMPLE FOOD PLANTS *Astragalus, Cirsium, Melilotus, Monarda, Penstemon, Phacelia, Salvia, Stachys, Trifolium, Vaccinium.*
BEHAVIOR Nests underground and less often on the surface. Males perch and chase moving objects in search of mates.
PARASITISM BY OTHER BEES Host to *B. insularis*, confirmed breeding record. It is also likely a host to *B. suckleyi*.

BOMBUS CITRINUS (SMITH, 1854)
LEMON CUCKOO BUMBLE BEE

LEFT: Female *Bombus citrinus*. JD
ABOVE: Male *Bombus citrinus*. GZ

IDENTIFICATION

Eastern species. Most similar to *B. variabilis* (see also *B. impatiens* and *B. bimaculatus*). The extreme light and dark female color patterns are rare.

HAND CHARACTERS Body size medium: female (no worker) 18–22 mm (0.69–0.87 inch). Hair of the metasoma short (but longer on metasomal T5 than length of last segment of the hind foot) and even. Hindleg tibia with the outer surface convex and densely hairy, lacking a pollen basket (contrast non-cuckoo bumble bees). Hair of the face black, usually with only a minority of yellow hairs intermixed, yellow hair above the base of the antenna rarely dense, *sides of the thorax entirely yellow* (contrast *B. variabilis*), upperside of the thorax often without black hair, but if present as a spot then not forming a band between the wings, metasomal T4–5 predominantly black (contrast cuckoo bumble bees other than *B. variabilis*). Wings light brown. Male 12–16 mm (0.48–0.64 inch). Eye similar in size and shape to the eye of any female bumble bee. Antenna of medium length, flagellum 3× longer than the scape. Hair of the face black with yellow hairs often intermixed above the antennal base, only rarely with a very few yellow hairs below the antennal base, upperside of the head yellow or with yellow hairs intermixed, *sides of the thorax yellow* (contrast *B. variabilis*), any black hairs confined to areas of the extreme underside and toward the back, metasomal *T4–7 predominantly black* (contrast all other cuckoo bumble bees) occasionally with a very few yellow hairs intermixed at the front and sides on T4.

MICROSCOPIC CHARACTERS Female metasomal T6 matte with dense pits or punctures (contrast *B. variabilis*), S6 with large keels that at the front are broadly rounded in section, weakly S-shaped in profile, the midline area with a broad weak third ridge. Male genitalia (similar to *B. insularis*) with the penis-valve head straight and about 4× longer than broad (contrast *B. bohemicus, B. suckleyi, B. flavidus*), the penis-valve angle on the underside of the shaft and to the side broadly triangular (contrast *B. bohemicus, B. suckleyi*), gonostylus inner arm with long branched hairs (contrast non-cuckoo bumble bees), volsella soft and yellow, broadly triangular (contrast *B. flavidus*), without inner hooks near the tip (contrast non-cuckoo bumble bees).

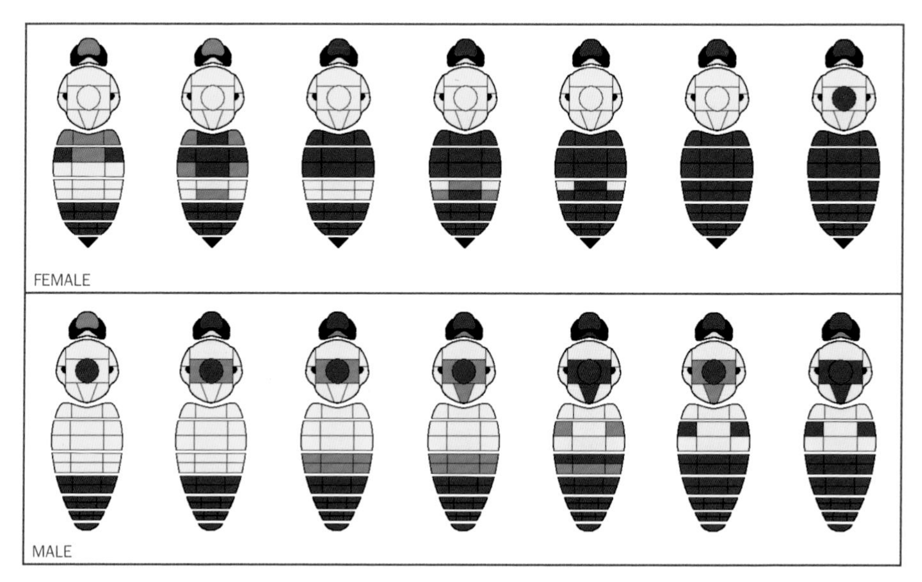

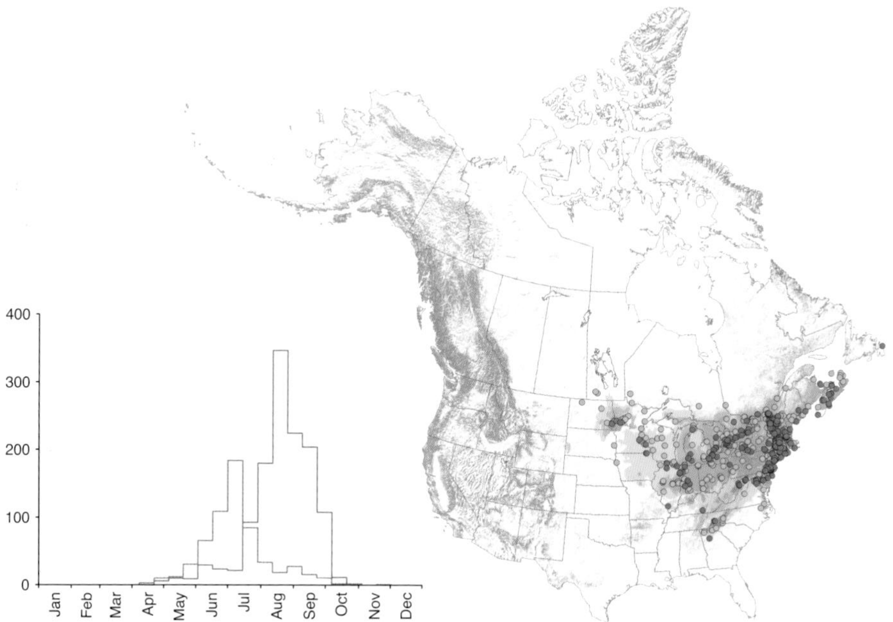

OCCURRENCE

RANGE AND STATUS Canadian Maritimes and eastern US in Eastern Temperate Forest and Boreal Forest regions, south in a narrow band at higher elevations along the Appalachian Mountains, west to the margin of the Great Plains.

EXAMPLE FOOD PLANTS *"Aster", Cirsium, Eupatorium, Liatris, Pycnanthemum, Solidago, Vernonia.*

BEHAVIOR Social parasite. Males patrol circuits in search of mates.

PARASITISM OF OTHER BEES This species is recorded as breeding as a parasite of colonies of *B. impatiens, B. bimaculatus,* and *B. vagans.*

BOMBUS VARIABILIS (CRESSON, 1872)
VARIABLE CUCKOO BUMBLE BEE

LEFT: **Female (posed) *Bombus variabilis*.** SCO
ABOVE: **Male *Bombus variabilis*.** VL

IDENTIFICATION

Eastern species. Most similar to *B. citrinus* (see also *B. impatiens, B. bimaculatus*). From morphology, *B. variabilis* appears to be very closely related to *B. intrudens* from Central America and might be part of the same species. The more extensively yellow female color pattern is rare.

HAND CHARACTERS Body size medium: female (no worker) 18–22 mm (0.73–0.87 inches). *Hair of the metasoma very short,* shorter on metasomal T5 than the length of the last segment of the hind foot (contrast other cuckoo bumble bees) and even. Hindleg tibia with the outer surface convex and densely hairy, lacking a pollen basket (contrast non-cuckoo bumble bees). Hair of the face black with only a minority of pale hairs above the base of the antenna, upperside of the thorax often with black hair, *sides of the thorax predominantly black* (contrast *B. citrinus*), hair of the metasoma entirely black (contrast cuckoo bumble bees other than *B. citrinus*). Wings dark brown. Male 15–17 mm (0.58–0.67 inches). Eye similar in size and shape to the eye of any female bumble bee. Antenna of medium length, flagellum 3× longer than the scape. Hair of the face black (contrast *B. insularis*), *sides of the thorax with many black hairs and metasomal T4–5 with many yellow hairs* (contrast *B. citrinus*). Hair on the metasoma very short and even, hair of T3–4 shorter than the hindleg basitarsus breadth (contrast *B. insularis*).

MICROSCOPIC CHARACTERS Female metasomal T6 shiny with sparse pits or punctures (contrast *B. citrinus*), S6 broadly triangular without a narrow spine at the tip (contrast *B. flavidus*), with large keels that are broadly rounded at the front in section, weakly S-shaped in profile, the midline area with a broad weak third ridge. Male genitalia with the penis-valve head straight and about 4× longer than broad (contrast *B. bohemicus, B. suckleyi, B. flavidus*), the penis-valve angle on the underside of the shaft and to the side broadly triangular (contrast *B. bohemicus, B. suckleyi*), gonostylus inner arm with long branched hairs (contrast non-cuckoo bumble bees), volsella soft and pale, broadly triangular (contrast *B. flavidus*), volsella maximum breadth viewed perpendicularly from above 1.5× the breadth of the far end of the gonocoxa (contrast *B. insularis*), without inner hooks near the tip (contrast non-cuckoo bumble bees).

OCCURRENCE

RANGE AND STATUS Eastern Temperate Forest and Great Plains region of the midwestern US, with scattered occurrences on the southeastern coastal plain, southern TX, and southern AZ. One of the rarest of all North American bumble bee species, it has been recorded only a few times in the

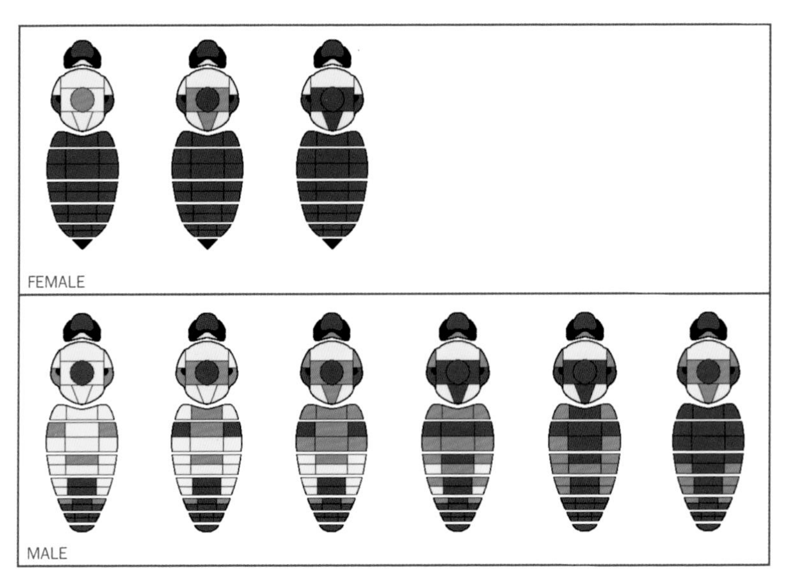

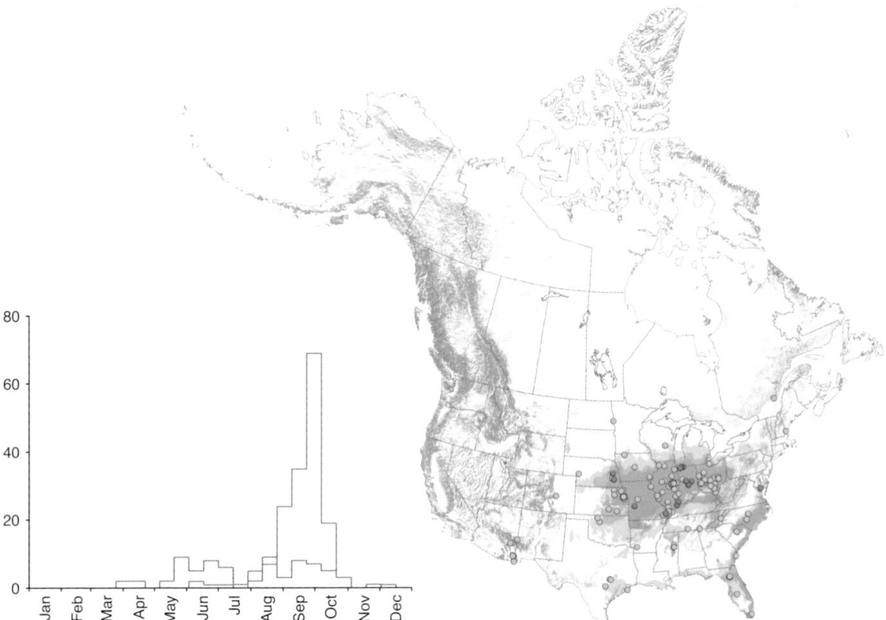

last 15 years. *B. variabilis* has always been uncommon in collections and has apparently declined along with its host, *B. pensylvanicus*.

EXAMPLE FOOD PLANTS *"Aster", Bidens, Echinacea, Helianthus, Solidago.*

BEHAVIOR Social parasite. Males are likely to patrol circuits in search of mates.

PARASITISM OF OTHER BEES This species is recorded as breeding as a parasite of colonies of *B. pensylvanicus*.

BOMBUS INSULARIS (SMITH, 1861)
INDISCRIMINATE CUCKOO BUMBLE BEE

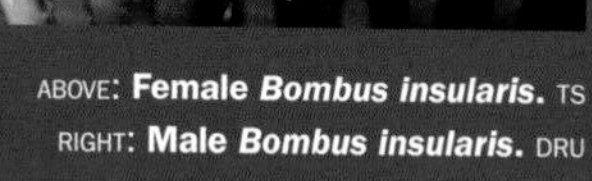

ABOVE: **Female *Bombus insularis*.** TS
RIGHT: **Male *Bombus insularis*.** DRU

IDENTIFICATION

Northern and western mountain species. Most similar to *B. suckleyi, B. bohemicus,* and *B. flavidus* (see also *B. vosnesenskii, B. caliginosus, B. vandykei, B. occidentalis, B. franklini,* and *B. flavidus*).

HAND CHARACTERS Body size medium: female (no worker) 16–20 mm (0.62–0.79 inch). Hair length medium. Hindleg tibia with the outer surface convex and densely hairy, lacking a pollen basket (contrast non-cuckoo bumble bees). *Hair of the face black with a dense yellow patch above the base of the antenna* (contrast *B. bohemicus, B. suckleyi, B. flavidus*), sometimes some yellow below the base of the antenna, but the face predominantly black, sides of the thorax yellow at the front but black on the underside and at the back, upperside of the thorax with the black band between the wings extending toward the back (scutellum) only narrowly along the midline between the yellow, metasomal T3 (in the East) usually with black hair along the midline and yellow at the sides or (in the west) sometimes entirely black, T4 at the sides extensively yellow. Wings light brown. Male 11–16 mm (0.44–0.62 inch). Eye similar in size and shape to the eye of any female bumble bee. Antenna of medium length, flagellum 3× longer than the scape. Hair of the face black with many yellow hairs above the base of the antenna (contrast *B. variabilis*) and only rarely completely absent below the antennal base, upperside of the head yellow or with yellow hairs intermixed, *metasomal T4 with many yellow hairs at the sides but a distinct patch of black hairs in the middle* (contrast *B. bohemicus, B. suckleyi, B. flavidus*), T7 black (contrast *B. flavidus*). Hair on the metasoma of medium length, hair of T3 at the front and in the middle longer than the hindleg basitarsus breadth (contrast *B. variabilis*).

MICROSCOPIC CHARACTERS Female metasomal S6 broadly triangular without a narrow spine at the tip (contrast *B. flavidus*), at the sides with large keels that at the front are broadly rounded in section, weakly S-shaped in profile, the midline area with a broad weak third ridge. Male genitalia (similar to *B. citrinus*) with penis-valve head straight and about 4× longer than broad (contrast

B. bohemicus, B. suckleyi, B. flavidus), the penis-valve angle of the shaft on the underside and to the side broadly triangular (contrast *B. bohemicus, B. suckleyi*), gonostylus inner arm with long branched hairs (contrast non-cuckoo bumble bees), volsella soft and yellow, broadly triangular (contrast *B. flavidus*), volsella maximum breadth viewed perpendicularly from above only just greater than the breadth of the far end of the gonocoxa (contrast *B. variabilis*), volsella without inner hooks near the tip (contrast non-cuckoo bumble bees).

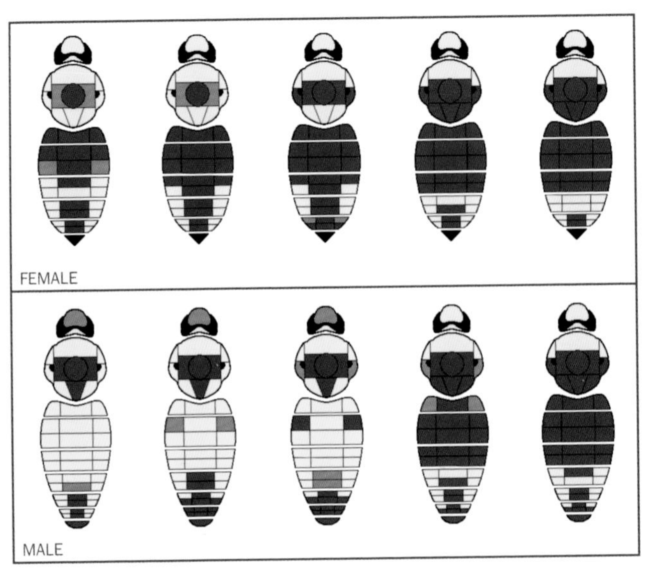

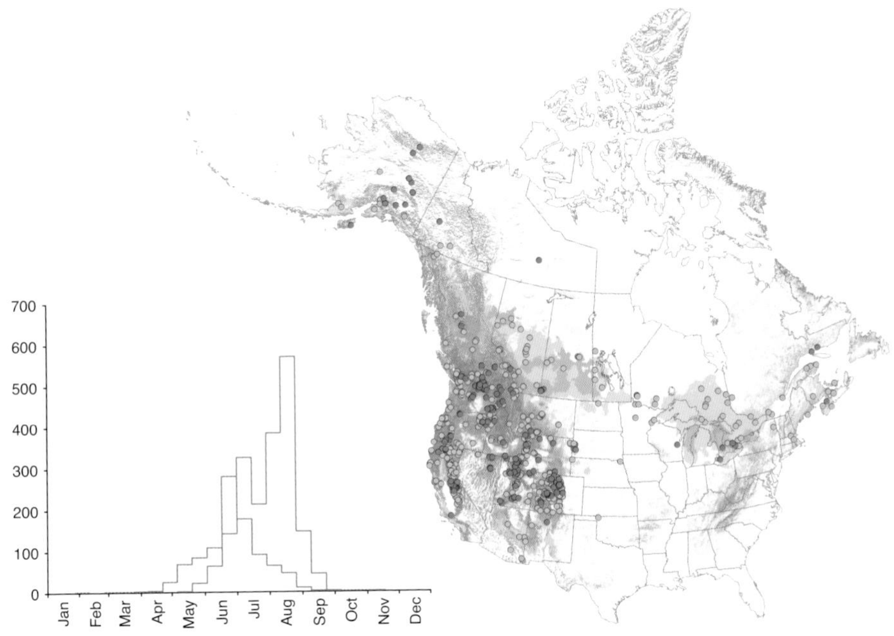

OCCURRENCE

RANGE AND STATUS Mountain West from AZ, NM north to AK, also scattered in the Tundra/Taiga region east to the Canadian Great Plains, ON Boreal Forest, and maritime northeastern US and eastern Canada. In decline in some parts of its range, possibly because of declines in host species.
EXAMPLE FOOD PLANTS *"Aster", Eupatorium, Heliomeris, Melilotus, Rubus, Senecio, Solidago, Trifolium, Vaccinium.*
BEHAVIOR Social parasite. Males patrol circuits in search of mates.
PARASITISM OF OTHER BEES This species is recorded as breeding as a parasite of colonies of *B. appositus, B. fervidus, B. flavifrons, B. nevadensis,* and *B. ternarius.* It has been recorded as present in colonies of *B. rufocinctus, B. nevadensis, B. occidentalis,* and *B. terricola.*

BOMBUS BOHEMICUS SEIDL, 1837
ASHTON CUCKOO BUMBLE BEE (INCLUDING *ASHTONI*)

Female *Bombus bohemicus.* RP

Male *Bombus bohemicus.* RW

IDENTIFICATION

Northern species. Most similar to *B. insularis, B. suckleyi,* and *B. flavidus.* Evidence from morphology and DNA barcodes supports a close relationship between *B. ashtoni* from North America and *B. bohemicus* from Europe and Asia, which appear to be parts of the same species.
HAND CHARACTERS Body size medium: female (no worker) 17–19 mm (0.65–0.73 inch). Hair of medium length. Hindleg tibia with the outer surface convex and densely hairy, lacking a pollen basket (contrast non-cuckoo bumble bees). Hair of the face and the upperside of the head black (contrast *B. insularis*), occasionally with some yellow hairs at the back of the upperside of the head, sides of the thorax predominantly black (contrast *B. suckleyi*), hair of metasomal T3–4 at the sides variable yellowish-white, but *T4 usually white at least in the middle at the back* (contrast *B. insularis, B. suckleyi, B. flavidus*). Male 11–17 mm (0.44–0.66 inch). Eye similar in size and shape to the eye of any female bumble bee. Antenna of medium length, flagellum 3× longer than

the scape. Hair of the hindleg basitarsus back fringe predominantly black (contrast *B. flavidus*), *sides of the thorax extensively black, metasomal T2 entirely black* (contrast *B. suckleyi*), T4 yellow without extensive black (contrast *B. insularis*), T7 black (contrast *B. flavidus*).

MICROSCOPIC CHARACTERS Female metasomal S6 broadly triangular and without a narrow spine at the tip (contrast *B. flavidus*), with large keels at the sides and near the front narrow and sharp in section, V-shaped in profile, strong but not protruberant so that they are scarcely visible when

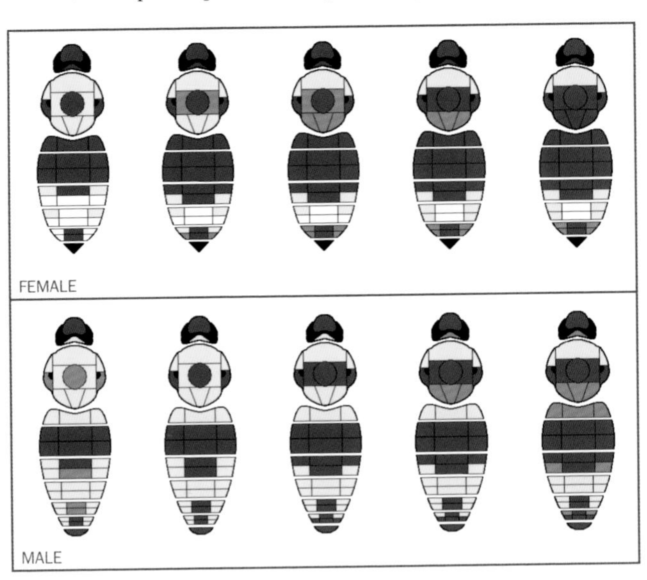

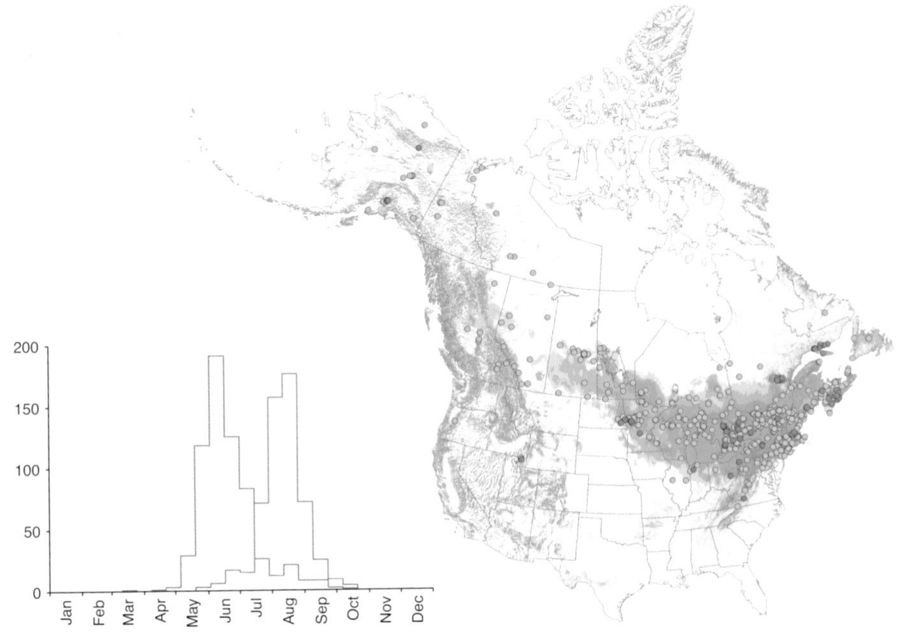

viewed from above as projecting beyond the sides of T6 (contrast *B. suckleyi*), the midline area rounded without a broad weak third ridge. Male genitalia with the penis-valve head straight and about 2× longer than broad (contrast *B. suckleyi, B. citrinus, B. variabilis, B. insularis*), the penis-valve angle on the underside of the shaft and to the side much reduced and not visible when viewed from above (contrast cuckoo bumble bees other than *B. suckleyi*), gonostylus inner arm with long branched hairs (contrast non-cuckoo bumble bees), volsella soft and yellow, broadly triangular (contrast *B. flavidus*), without inner hooks near the tip (contrast non-cuckoo bumble bees).

OCCURRENCE

RANGE AND STATUS Eastern and midwestern US and Canada in Eastern Temperate Forest and Boreal Forest regions, south in a narrow band at higher elevations along the Appalachian Mountains, and extending northwest through the Canadian Great Plains, Mountain West, and Tundra/Taiga to AK. Apparently declining rapidly throughout its large range, presumably linked to declines of some of its hosts. Also in Europe and Asia.

EXAMPLE FOOD PLANTS *Cirsium, Melilotus, Rubus, Rudbeckia, Solidago, Symphyotrichum, Trifolium, Vaccinium.*

BEHAVIOR Social parasite. Males patrol circuits in search of mates.

PARASITISM OF OTHER BEES This species is recorded as breeding as a parasite of colonies of *B. terricola* and *B. affinis*. It is likely that it also breeds in colonies of *B. occidentalis* and possibly *B. cryptarum*.

BOMBUS SUCKLEYI GREENE, 1860
SUCKLEY CUCKOO BUMBLE BEE

LEFT: **Female *Bombus suckleyi*.** CB
ABOVE: **Male *Bombus suckleyi*.** AR

IDENTIFICATION

Western species. Most similar to *B. bohemicus, B. insularis,* and *B. flavidus.*

HAND CHARACTERS Body size medium: female (no worker) 18–23 mm (0.72–0.92 inch). Hair short and even. Hindleg tibia with the outer surface convex and densely hairy, lacking a pollen basket (contrast non-cuckoo bumble bees). Hair of the face black (contrast *B. insularis*), sides of the thorax usually predominantly yellow (contrast *B. bohemicus*), metasomal T4 at the front black and with black continuously along the midline (contrast *B. bohemicus*). Male 13–16 mm (0.51–0.61 inch). Eye similar in size and shape to the eye of any female bumble bee. Antenna of medium length, flagellum 3× longer than the scape. Hair of the sides of the thorax yellow, with at most black hairs only toward the back, metasomal T2 extensively yellow (contrast *B. bohemicus*), T4 yellow with at most only a very narrow line of black hairs along the midline (contrast *B. insularis*), T7 black (contrast *B. flavidus*).

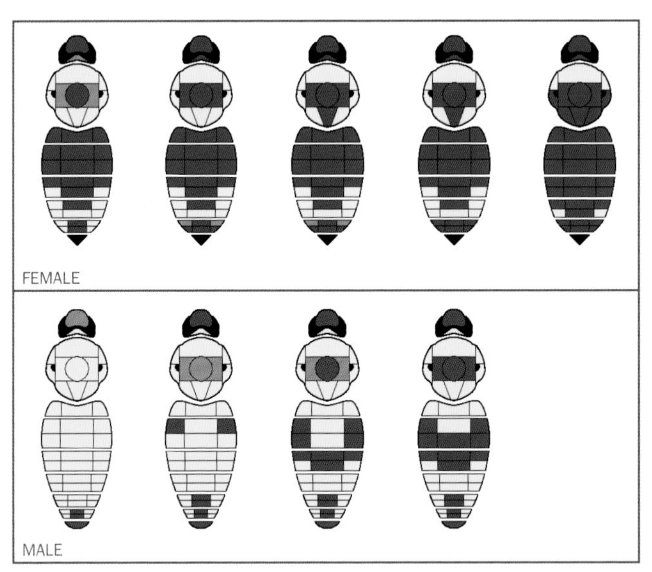

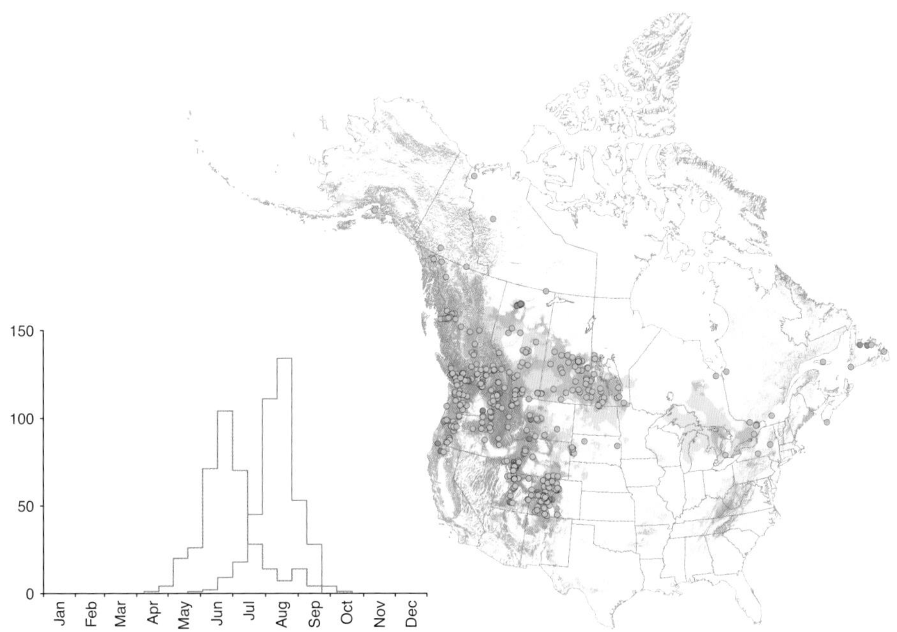

MICROSCOPIC CHARACTERS Female metasomal S6 broadly triangular and without a narrow spine at the tip (contrast *B. flavidus*), with large keels at the sides and near the front narrow and sharp in section, V-shaped in profile and *protuberant, so that they are clearly visible when viewed from above as projecting beyond the sides of T6* (contrast *B. bohemicus*), the midline area rounded without a broad weak third ridge. Male genitalia with the penis-valve head straight and just less than 3× longer than broad (contrast *B. bohemicus, B. citrinus, B. variabilis, B. insularis*), the penis-valve angle on the underside of the shaft and to the side much reduced and not visible from above (contrast cuckoo bumble bees other than *B. bohemicus*), gonostylus inner arm with long branched hairs (contrast non-cuckoo bumble bees), volsella soft and yellow, broadly triangular (contrast *B. flavidus*), without inner hooks near the tip (contrast non-cuckoo bumble bees).

OCCURRENCE

RANGE AND STATUS Mountain West from CA and CO to AK, east to the Canadian Great Plains, with a disjunct population in NL. Apparently declining in some parts of its range.
EXAMPLE FOOD PLANTS *"Aster", Chrysothamnus, Cirsium, Solidago.*
BEHAVIOR Social parasite. Males patrol circuits in search of mates.
PARASITISM OF OTHER BEES This species is recorded as breeding as a parasite of colonies of *B. occidentalis*. It has been recorded as present in colonies of *B. terricola, B. rufocinctus, B. fervidus, B. nevadensis,* and *B. appositus*.

BOMBUS FLAVIDUS EVERSMANN, 1852
FERNALD CUCKOO BUMBLE BEE (INCLUDING *FERNALDAE*)

Female *Bombus flavidus*. JG

Male *Bombus flavidus*. DRU

IDENTIFICATION

Widespread species. Most similar to *B. insularis*, *B. bohemicus*, and *B. suckleyi* (see also *B. vosnesenskii, B. caliginosus, B. vandykei, B. occidentalis,* and *B. franklini*). Evidence from morphology and from DNA barcodes supports a close relationship between *B. fernaldae* from North America and *B. flavidus* from Europe and Asia, which appear to be parts of the same species. The darker color patterns are confined to the west.

HAND CHARACTERS Body size small: female (no worker) 17–18 mm (0.66–0.72 inch). Hair of medium length. Hindleg tibia with the outer surface convex and densely hairy, lacking a pollen basket (contrast non-cuckoo bumble bees). Hair of the face black with at most a few yellow hairs above the antennal base, metasomal T4 mostly yellow with at the front a small patch of black hairs. Male 11–15 mm (0.44–0.59 inch). Eye similar in size and shape to the eye of any female bumble bee. Antenna of medium length, flagellum 3× longer than the scape. Hair of the face black,

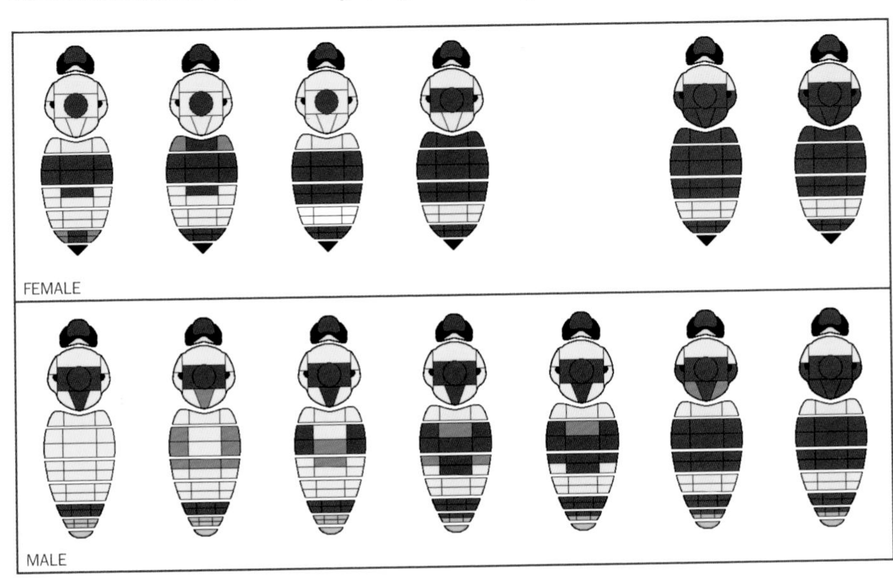

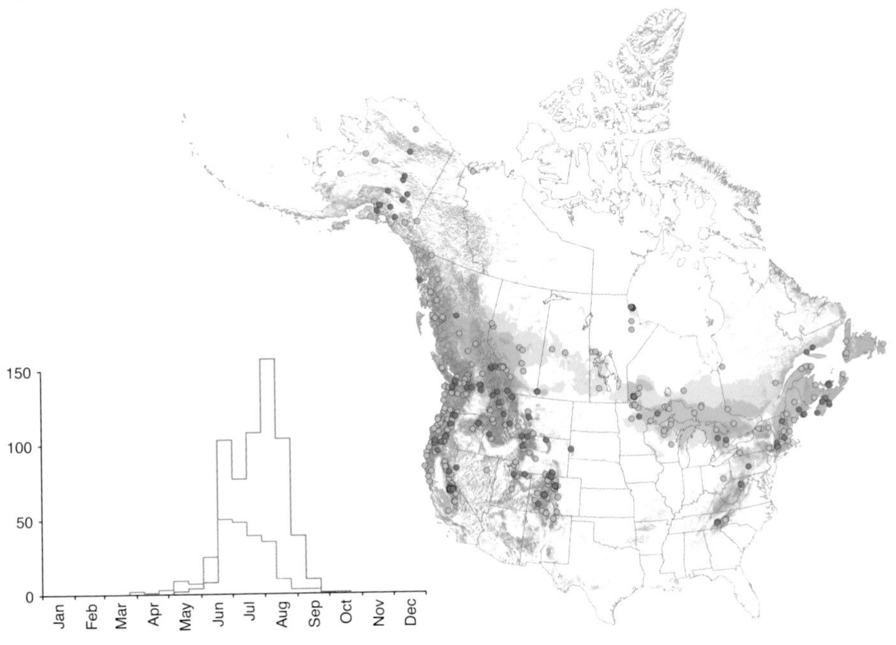

hindleg basitarsus with the back fringe predominantly yellow (contrast *B. bohemicus*), metasomal T1–3 sometimes nearly completely yellow, T4 yellow with at most a few black hairs near the midline (contrast *B. insularis*), *T6–7 with orange hairs that often have paler tips* (contrast *B. insularis, B. bohemicus, B. suckleyi*).

MICROSCOPIC CHARACTERS Female metasomal *T6 curled very strongly under the metasoma and pointing to the front, S6 with the keels at the sides small, the tip narrowed and terminating in a broad shiny spine* (contrast *B. insularis, B. bohemicus, B. suckleyi*). Male genitalia with the penis-valve head straight and less than 3× longer than broad, the penis-valve angle on the underside of the shaft and to the side broadly triangular (contrast *B. bohemicus, B. suckleyi*), gonostylus inner arm with long branched hairs (contrast non-cuckoo bumble bees), volsella soft and yellow and *long, narrow, and shaped like a finger* (contrast *B. insularis, B. bohemicus, B. suckleyi,* other cuckoo bumble bees), without inner hooks near the tip (contrast non-cuckoo bumble bees).

OCCURRENCE

RANGE AND STATUS Widely scattered across the continent, from the northeastern US and adjacent Canada in the Boreal Forest region, less commonly in Eastern Temperate Forest, south in a narrow band at higher elevations along the Appalachian Mountains, sparingly in the northern Great Plains of Canada, in the Mountain West south to CO, NM, and on the Pacific Coast from CA north to AK, with scattered records in the Tundra/Taiga of Canada. Also in Europe and Asia.

EXAMPLE FOOD PLANTS *Cirsium, Heliomeris, Melilotus, Senecio.*

BEHAVIOR Social parasite. Males patrol circuits in search of mates.

PARASITISM OF OTHER BEES In Europe, recorded as breeding as a parasite of colonies of bumble bee species of the subgenus *Pyrobombus*. It is also the most common cuckoo species found in nests of *Pyrobombus* in North America, but surprisingly there are no direct records of it breeding in host colonies of any species from North America. It has also been recorded as present in colonies of *B. rufocinctus, B. occidentalis,* and *B. appositus.*

IDENTIFICATION KEYS TO FEMALE AND MALE BUMBLE BEES, WITH PHOTOS

Numbers in parentheses refer to the number of the statement in a couplet from which a couplet is reached. In the following keys, the term *cheek* refers to the oculo-malar area.

— FEMALES

1a Hindleg tibia with a pollen basket (corbicula), the outer surface flat without long hair in the center but with long anterior and posterior fringes (Fig. 1); S6 without lateral keels → **2**

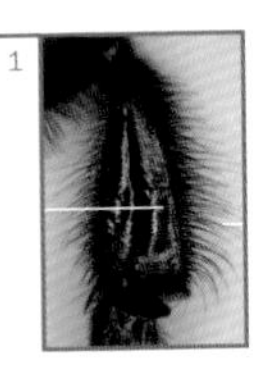

FIG 1

1b Hindleg tibia without a pollen basket, the outer surface convex with dense long hair in the center as well as short anterior and posterior fringes (Fig. 2); S6 with lateral keels → **42 (Group 4 *Psithyrus*)**

FIG 2

2a (1a) Midleg basitarsus distal (furthest from the body) posterior corner rounded (>45°) (Fig. 3) → **3**

2b Midleg basitarsus distal posterior corner with a sharp spine (<45°) (Fig. 4) → **36 (Group 3)**

FIG 3 FIG 4

3a (2a) Cheek about as long (shortest distance from the edge of the eye to the edge of the cheek between the mandibular hinges) as broad (distance between and including the mandibular hinges) or longer than broad (Fig. 5), the lateral ocellus always small (so at least twice the ocellar diameter from the edge of the eye) and its center located posterior to the narrowest (anterior) line between the eyes (Fig. 6) → **4 (Group 1)**

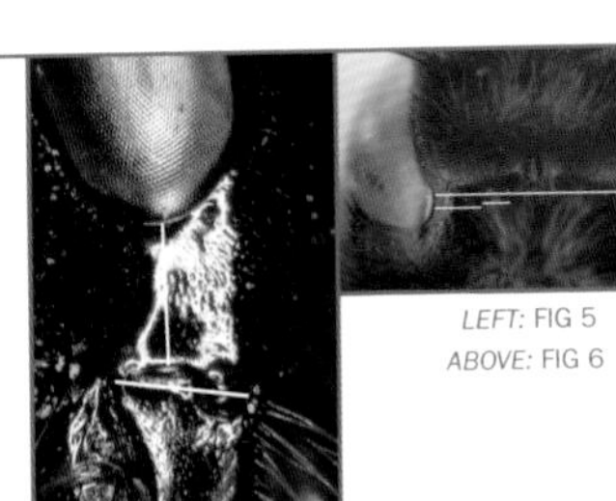

LEFT: FIG 5
ABOVE: FIG 6

3b Cheek shorter than broad (Fig. 7), the lateral ocellus small (so more than twice the ocellar diameter from the edge of the eye) *and* its center located posterior to the narrowest line between the eyes, *or if* the cheek is nearly equal in length

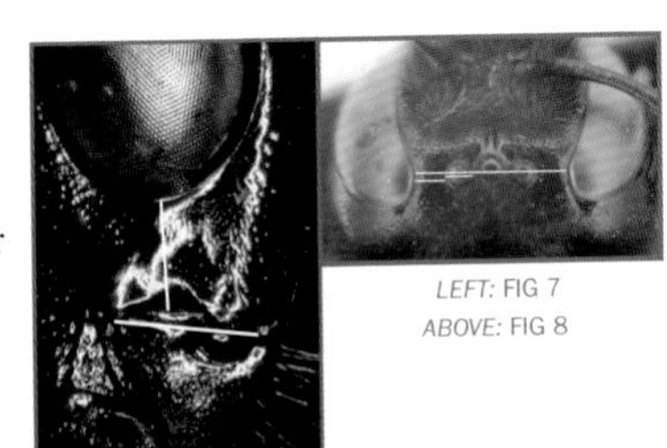

LEFT: FIG 7
ABOVE: FIG 8

and breadth *then* the lateral ocellus is large (so <1.8× the ocellar diameter from the edge of the eye) *and* its center located on the narrowest (anterior) line between the eyes (Fig. 8) → **27 (Group 2)**

4a (3a) **Group 1**
Cheek approximately square or just longer than broad (Fig. 9), mandible with a very shallow notch (incisura) anterior to the tooth at the posterior distal corner, the depth of the notch less than a third of its width and often scarcely perceptible (Fig. 10), inner eye margin opposite the lateral ocellus with a band of large pits or punctures, the punctures spaced by more than their own widths, and the areas between the large punctures flat and shining with very few or no small punctures so that the band appears sparse and shining (Fig. 11) → **5 (*Pyrobombus*)**

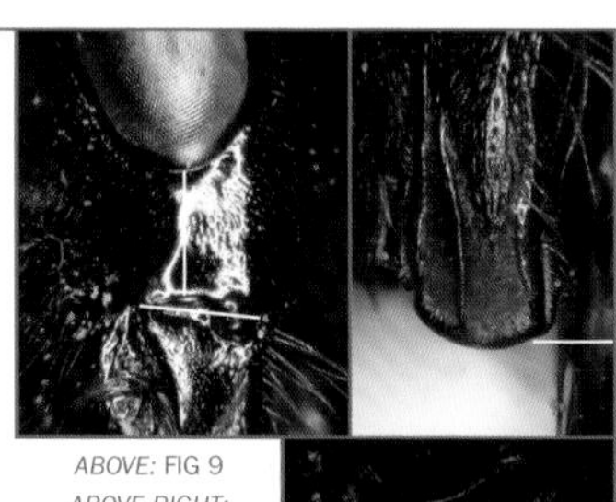

ABOVE: FIG 9
ABOVE RIGHT: FIG 10

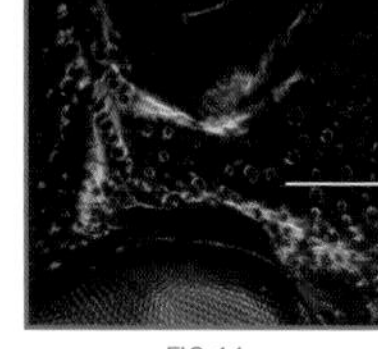

FIG 11

4b *Either* the cheek much longer than broad, *or if* the cheek is only slightly longer than broad (Fig. 12) *then* the mandible has a deep notch (incisura) anterior to the tooth at the posterior distal corner, the depth of the notch usually more than a third its width (Fig. 13) *and also* the inner eye margin opposite the lateral ocellus has a band of large and small pits or punctures, many punctures spaced by less than their own widths, so that the area appears moderately dense and matte (Fig. 14) → **24 (*Alpinobombus*)**

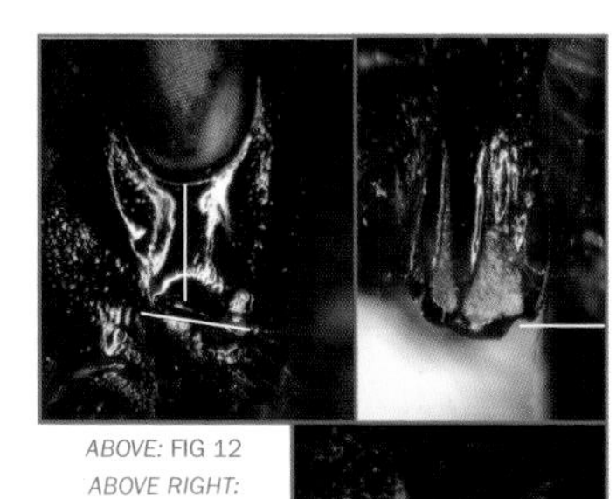

ABOVE: FIG 12
ABOVE RIGHT: FIG 13

FIG 14

5a (4a) *Pyrobombus*
Hair of T1–2 entirely black, or very rarely T1–2 in the posterior lateral corners with small yellow patches → **6**

5b Hair of T1 and often T2 extensively yellow or cream-white → **8**

6a (5a) Hair of T3 yellow at least in the posterior half, T4 often black → ***B. vandykei***

6b Hair of T3 entirely black, T4 mostly yellow → **7**

7a (6b) Hair of T4 almost entirely yellow with just a few black hairs on the midline, S3–4 with posterior hair fringes black; cheek as long as broad → ***B. vosnesenskii***

7b Hair of T4 yellow but with black anteriorly, often forming a triangle in the middle and interrupting the yellow along the midline, S4 or S3–4 with posterior hair fringes usually yellow laterally; cheek slightly longer than broad → ***B. caliginosus***

8a (5b) Cheek about as long as broad (Fig. 15); (many color patterns, but includes almost all individuals with the hair of T2 entirely black, all individuals with either T2 or T5 at least partly orange or red or pink [may fade to nearly yellow], and all individuals with T4 white) → **9**

FIG 15

8b Cheek clearly slightly longer than broad (Fig. 16); (many color patterns, but the hair of T2 usually has at least some yellow, only very rarely entirely black, *or if* there is orange-red hair on the metasoma *then* both T2 and T5 are a similar yellow and/or black, and T4 never white) → **19**

FIG 16

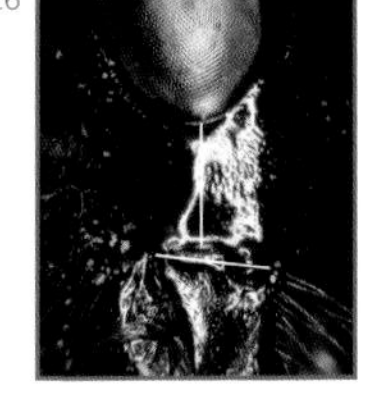

9a (8a) Hair of T5 mostly orange or pink or brown, *or if* T5 is mostly yellow *then* pale hairs predominate medially and the thoracic dorsum anteriorly has many black hairs intermixed throughout the yellow or white → **10**

9b Hair of T5 mostly black or white, *or if* T5 is mostly yellow or brown *then either* black hairs predominate along the midline *or* the thoracic dorsum anteriorly has few or no black hairs intermixed with the yellow near the midline → **12**

10a (9a) Hair of the thoracic dorsum yellow with a sharply defined black band between the wings → ***B. frigidus***

10b Hair of the thoracic dorsum yellow or gray intermixed with black, any darker area between the wings weakly defined → **11**

11a (10b) Hair of the thoracic dorsum with as much or almost as much pale hair intermixed posteriorly as anteriorly, forming subtly lighter bands anteriorly and posteriorly, T2 medially yellow and anteriorly without black, although T2 sometimes laterally and posteriorly black, T5 pale orange or pink or rarely yellow → ***B. mixtus***

11b Hair of the thoracic dorsum with much less pale hair intermixed posteriorly than anteriorly, forming a subtly lighter band only anteriorly, T2 yellow sometimes narrowly interrupted posteriorly and medially by black or with black hairs intermixed, T5 pale brownish yellow → ***B. sitkensis***

12a (9b) Hair of T2–6 entirely black → ***B. impatiens***

12b Hair of T2–6 with some yellow or orange → **13**

13a (12b) Hair of T2 almost entirely yellow, with only a few scattered black hairs laterally, T4 black → ***B. sandersoni***

13b Hair of T2 often orange, black, or part yellow, or *if* T2 is entirely yellow *then* T4 white → **14**

14a (13b) Hair of T4–5 white → ***B. jonellus***

14b Hair of T4–5 yellow or black → **15**

15a (14b) Hair of the thoracic dorsum with the posterior yellow or gray band (scutellum) entirely divided along the midline by black → **16**

15b Hair of the thoracic dorsum with the posterior yellow band not entirely divided along the midline by black → **17**

16a (15a) Hindleg tibia with the long hairs of the corbicular fringes black, T2 orange with at most a few black hairs anteriorly in the middle; hindleg basitarsus surface black like the tibia → ***B. ternarius***

16b Hindleg tibia with the long hairs of the corbicular fringes usually mostly pale orange or brown, *or if* T2 is black *then* the corbicular fringes are sometimes nearly black, T2 usually with at least an obvious large triangle of black hair anteriorly in the middle; hindleg basitarsus surface brown and lighter than the black tibia → ***B. bifarius***

17a (15b) Hair of the face and head dorsally mostly yellow with only a few black hairs, thoracic dorsum posteriorly entirely yellow → ***B. huntii***

17b Hair of the face and head dorsally with many black hairs, thoracic dorsum posteriorly yellow with some black hairs either intermixed or along the midline → **18**

18a (17b) Hair of the face, head dorsally, and thoracic dorsum anteriorly yellow with many black hairs intermixed, posteriorly (scutellum) with fewer black hairs intermixed than anteriorly, T2 either orange or black without yellow → ***B. melanopygus***

18b Hair of the face black with a small patch of yellow at the base of the antenna, head dorsally and thoracic dorsum anteriorly yellow with few black hairs intermixed, posteriorly (scutellum) with black hairs only medially and anteriorly, T2 with orange and/or black and usually with some yellow anteriorly near the midline → ***B. sylvicola***

19a (8b) Hair of the thoracic dorsal anterior yellow band with at most very few black hairs → **20**

19b Hair of the thoracic dorsal anterior yellow or gray band with many black hairs intermixed → **23**

20a (19a) Hair of T3–4 orange-red (sometimes faded) → ***B. centralis***

20b Hair of T3–4 black or yellow → **21**

21a (20b) Hair of the thoracic dorsum yellow or brown with only a few inconspicuous black hairs between the wings, side of the thorax extensively black, or *if* yellow *then* T3 with yellow near the middle, T6 posteriorly and medially with a broad rounded raised bump (boss) just before the apex → ***B. perplexus***

21b Hair of the thoracic dorsum with black intermixed conspicuously between the wings, side of the thorax yellow, T3 black, T6 posteriorly and medially without a broad rounded raised bump just before the apex → **22**

22a (21b) Hair of the face black or with only a few yellow hairs intermixed, T2 usually extensively black at least in the anterior lateral corners, the medial yellow area often forming a W shape → ***B. bimaculatus***

22b Hair of the face with shorter yellow hairs often broadly intermixed, T2 anterior margin usually entirely yellow (unless the yellow band is very narrow) and the posterior margin yellow or with black → ***B. vagans***

23a (19b) Hind tibia with long corbicular fringes black, T3–4 laterally and S3–5 black with few or no yellow hairs → ***B. vagans***

23b Hind tibia with long corbicular fringes often at least in part orange-brown, T3–4 laterally and S3–5 mostly yellow → ***B. flavifrons***

24a (4b) *Alpinobombus*
Cheek just longer than broad (Fig. 17), hindleg tibia outer (corbicular) surface coarsely rough and matte with the microsculpture (surface texture) interrupting the reflective highlights (Fig. 18); hair of the thoracic dorsum sometimes entirely black, or often with anterior and posterior yellow bands,

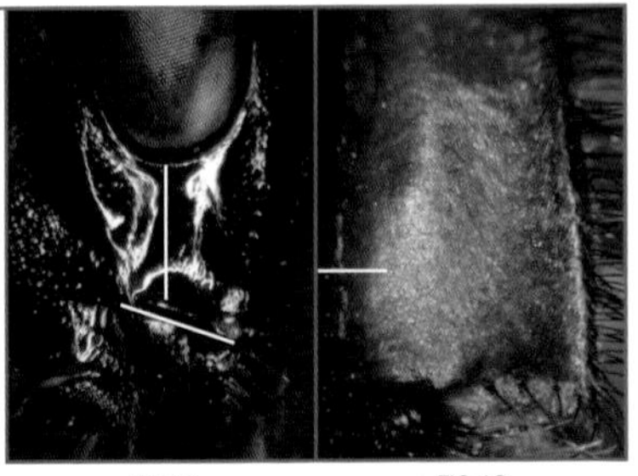

FIG 17 FIG 18

if T3 has some yellow *then* this is concentrated in a central patch near the midline and the posterior edge, *or if* T3–6 are mostly black *then* the thoracic dorsal anterior yellow band is often nearly twice the breadth of the posterior band → ***B. polaris***

24b Cheek much longer than broad (Fig. 19), hindleg tibia outer surface smooth and shining with the microsculpture scarcely or not interrupting the reflective highlights (Fig. 20); hair of the thoracic dorsum always with an anterior yellow band and usually with a posterior yellow band, or rarely with many black hairs intermixed, *if* T3 has some yellow *then* this is either predominant throughout T3 or concentrated near the lateral and posterior edges with less near the midline, *or if* T3–6 are mostly black *then* the thoracic dorsal anterior yellow band is nearly equal in breadth to the posterior band → **25**

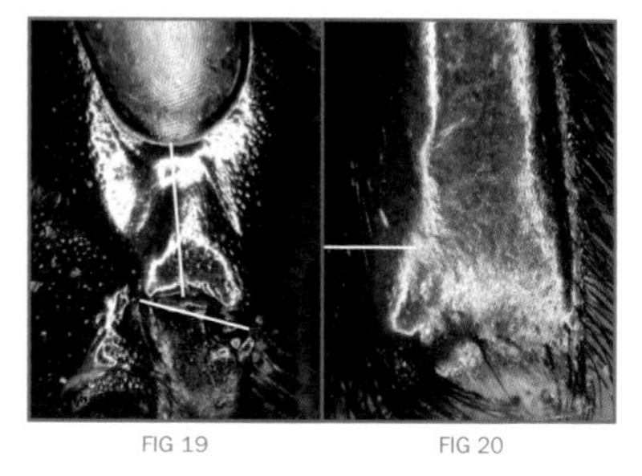

FIG 19 FIG 20

25a (24b) Hair of the face black but often with a tuft of yellow at the antennal base, side of the thorax throughout at least the dorsal half yellow and often with only a few black hairs ventrally and posteriorly, very rarely with black intermixed, T3 usually with at least traces of yellow as a lateral longitudinal fringe that continues around as a posterior fringe → ***B. balteatus***

25b Hair of the face black, side of the thorax black at least in the ventral half, or entirely, or with black intermixed, T3 almost entirely yellow or entirely black → **26**

26a (25b) Hair of T3–5 entirely black, sometimes with small pale tips, thoracic dorsum posterior to the wings (scutellum) yellow → ***B. hyperboreus***

26b Hair of T3 usually almost completely yellow, or *if* T3 black *then* the thoracic dorsum posterior to the wings is also mostly black,T5 often with orange → ***B. neoboreus***

27a (3b) **Group 2**
Mandible with a shallow notch (incisura) anterior to the tooth at the posterior distal corner (Fig. 21), lateral ocellus large, so that it is less than 1.8× the ocellar diameter from the edge of the eye, with its center located on the narrowest (anterior) line between the eyes (Fig. 22) → **32 (*Cullumanobombus*)**

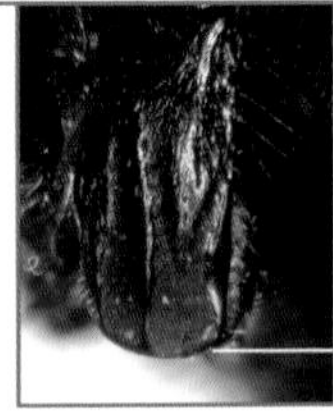

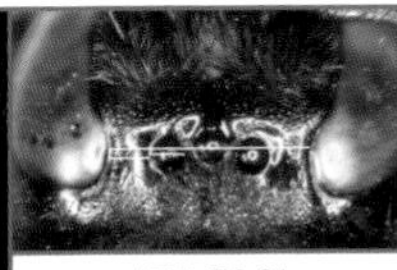

LEFT: FIG 21
ABOVE: FIG 22

27b Mandible with a deep notch anterior to the tooth at the posterior distal corner (Fig. 23), lateral ocellus small, so that it is more than twice the ocellar diameter from the edge of the eye, with its center located strongly posterior to the narrowest (anterior) line between the eyes (Fig. 24) → **28 (*Bombus*)**

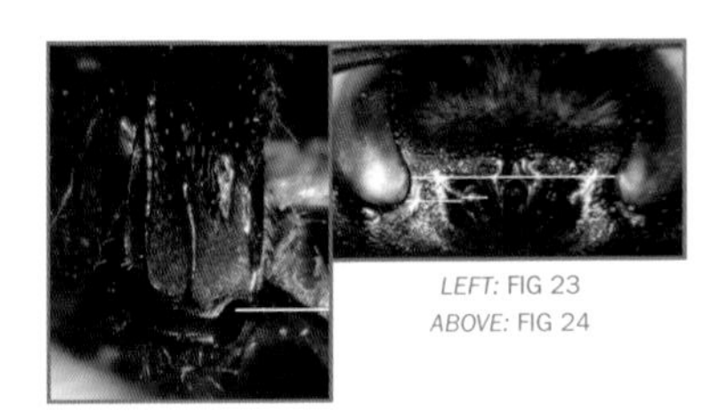

LEFT: FIG 23
ABOVE: FIG 24

28a (27b) *Bombus s. str.*
Hair of the side of the thorax yellow, T2 yellow or often with some brown, T3–5 entirely black → ***B. affinis***

28b Hair of the side of the thorax with some black, some of T3–5 with some pale hair → **29**

29a (28b) Hair of the thoracic dorsum between the wings bright yellow with at most a small patch of black at the center, thoracic dorsum posteriorly (scutellum) and T1–4 black; hair very short and even → ***B. franklini***

29b Hair of the thoracic dorsum between the wings with black, some of T1–4 with pale hair; hair short, medium, or long → **30**

30a (29b) Hair of T2 mostly yellow, T3 black → ***B. cryptarum***

30b Hair of T2 black, or *if* T2 yellow *then* T3 yellow → **31**

31a (30b) *Either* the hair of the thoracic dorsum between the wings black *and* T2 entirely yellow, *or* T2 anteriorly with a few black hairs *and* T5 almost entirely black → ***B. terricola***

31b *Either* the hair or the thoracic dorsum between the wings with some yellow *and* T3 yellow with a few black hairs anteriorly, *or* T2 anteriorly black *and* T5 with pale hair → ***B. occidentalis***

32a (27a) *Cullumanobombus*
Cheek very much shorter (<0.5×) than broad (Fig. 25); hair of the head dorsally black, T1–2 entirely yellow, and T3–5 entirely black; hair of T3 lying completely flat to the body surface → ***B. fraternus***

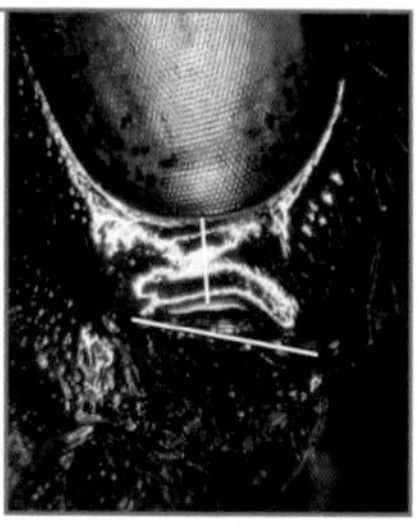

FIG 25

32b Cheek shorter (≥0.5×) than broad (Fig. 26); hair of the head dorsally yellow, and/or T1–2 partly black, and/or T3–5 partly yellow; hair of T3 erect and rising clear of the body surface → **33**

FIG 26

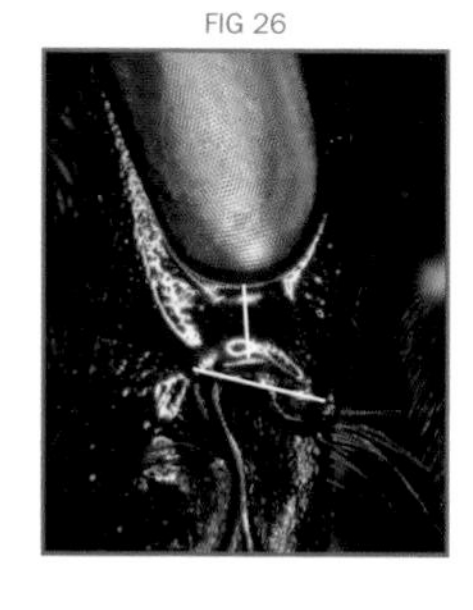

33a (32b) Cheek much shorter (about 0.5×) than broad (Fig. 27), inner eye margin dorsally opposite the lateral ocellus with pits or punctures occupying a band consistently broader than half the distance to the lateral ocellus, the punctures mostly small and spaced by less than their own widths, the surface appearing matte (Fig. 28) → ***B. rufocinctus***

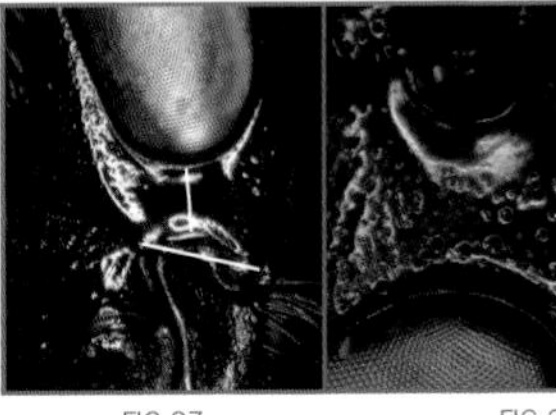
FIG 27 FIG 28

33b Cheek a little shorter than (>0.6× as long as) broad (Fig. 29), inner eye margin dorsally opposite the lateral ocellus with a few pits or punctures occupying a band in part less than half the distance to the lateral ocellus, for the narrow part of the band's length the punctures mostly large and spaced by more than their own widths with largely smooth areas between them (Fig. 30) → **34**

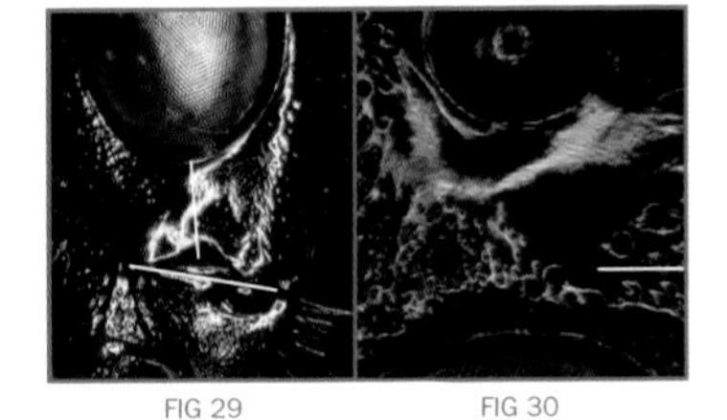
FIG 29 FIG 30

34a (33b) Hair of the thoracic dorsum and T1 black except for a yellow band across the anterior of the thorax, T5 black or sometimes orange → ***B. crotchii***

34b Hair of the thoracic dorsum and T1 mostly yellow, T5 black → **35**

35a (34b) Hair of the side of the thorax yellow, thoracic dorsum often with a black spot, head dorsally and T2 usually at least part black, T2 sometimes part brown, T3 black → ***B. griseocollis***

35b Hair of the side of the thorax black, thoracic dorsum without black, head dorsally and T2 yellow, T3 at least part yellow → ***B. morrisoni***

36a **Group 3**
(2b) Mandible with the anterior keel reaching the distal margin (Fig. 31), hindleg basitarsus with the proximal (closest to the body) posterior process pointed and longer than broad (Fig. 32) → **37**

FIG 31

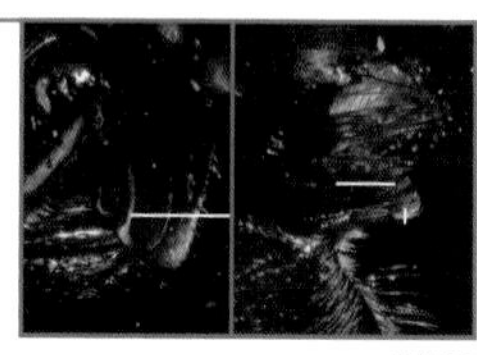

FIG 32

36b Mandible with the anterior keel not reaching the distal margin (Fig. 33), hindleg basitarsus with the proximal posterior process blunt and shorter than broad (Fig. 34) → **41** **(*Bombias*)**

FIG 33

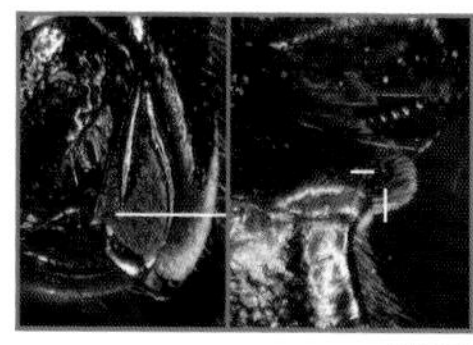

FIG 34

37a Clypeus surface rough with many large
(36a) and small pits or punctures (Fig. 35); hair of the face black → **38 (*Thoracobombus*)**

FIG 35

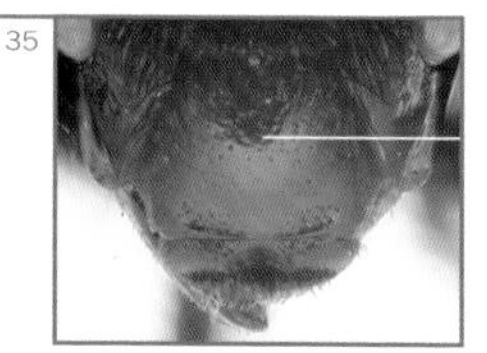

37b Clypeus surface very smooth and shiny with only few very small pits or punctures near the center (Fig. 36); hair of the face usually yellow or gray-white, only rarely black → **39** **(*Subterraneobombus*)**

FIG 36

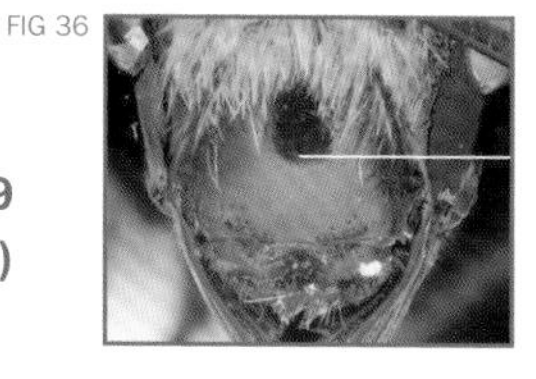

38a *Thoracobombus*
(37a) Hair of the side of the thorax either yellow, or *if* black *then* T2 also black → ***B. fervidus***

38b Hair of the side of the thorax mostly black, T2 yellow → ***B. pensylvanicus***

39a *Subterraneobombus*
(37b) Hair of the face mostly black at least above the antennal base, *if* there is a patch of yellow below the antennal base *then* it is a shade of yellow that closely matches T1, T5 mostly yellow but often with some black hairs anteriorly → ***B. distinguendus***

39b Hair of the face and the anterior thoracic dorsum a gray-white or cream-yellow that is usually paler than T1, T5 yellow or orange-brown or black, or *if* the face has many black hairs intermixed *then* T5 is mostly black → **40**

40a (39b) Hair of the face usually cream-yellow or rarely with many black hairs intermixed, T1 a darker golden yellow, T5 usually mostly black → ***B. borealis***

40b Hair of the face usually gray-white, T1 and T5 orange-brown or straw yellow, or *if* the face is straw yellow *then* T1 and T5 are also straw yellow → ***B. appositus***

41a *Bombias*
(36b) Inner eye margin dorsally opposite the lateral ocellus with a band of large pits or punctures with the spaces between them rough and dull (Fig. 37); hair of T5 black *and* between the wings dorsally with a strong black band, without yellow hairs → ***B. auricomus***

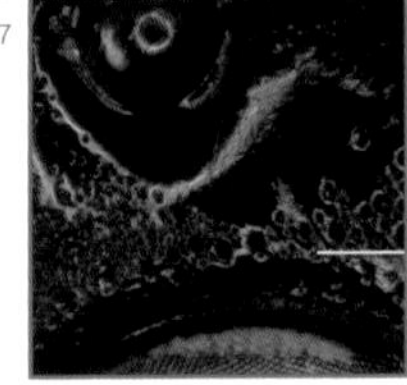
FIG 37

41b Inner eye margin dorsally opposite the lateral ocellus with a band of large pits or punctures with the spaces between them smooth and shining (Fig. 38); *either* hair of T5 orange *or* between the wings dorsally with yellow strongly intermixed within the black band → ***B. nevadensis***

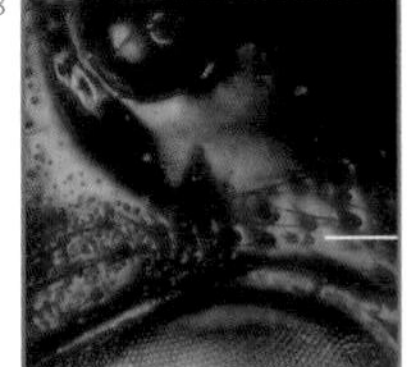
FIG 38

42a **Group 4** *Psithyrus*
(1b) T6 very strongly curled under the metasoma so that the apex points anteriorly (Fig. 39), S6 strongly projecting distally (at the tip) beyond T6 as a shiny curved spine, the lateral keels just small bumps (Fig. 40) → ***B. flavidus***

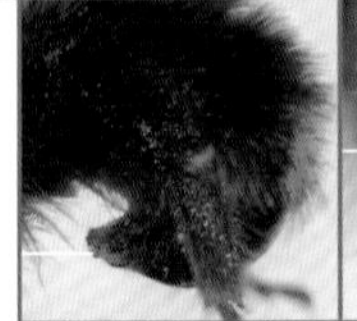
FIG 39

FIG 40

42b T6 at most weakly curved under the metasoma so that the apex points ventrally (Fig. 41), S6 scarcely projecting distally (at the tip) beyond T6, the lateral keels long and strongly projecting (Fig. 42) → **43**

FIG 41

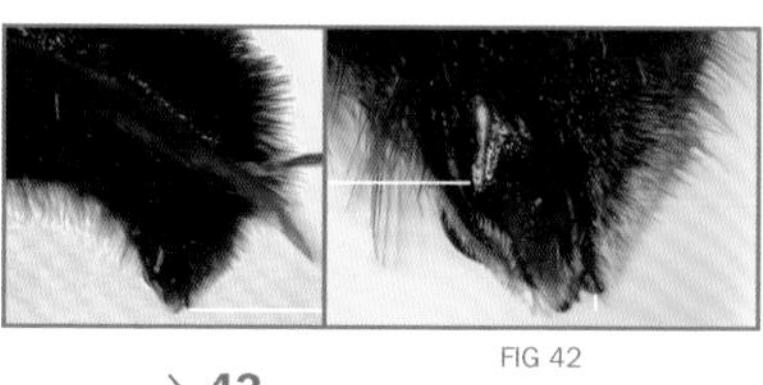

FIG 42

43a (42b) S6 with the lateral keels anteriorly evenly rounded in section, weakly S-shaped in profile, posterior half narrow, median area between them with a weakly angled third longitudinal ridge (Fig. 43); hair of T4 black, or *if* T4 yellow *then* the face above the antennal base mostly yellow → **44**

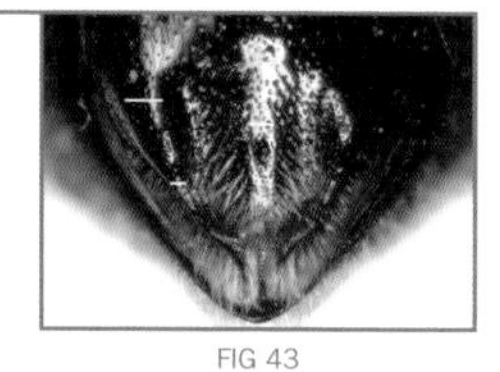

FIG 43

43b S6 with the lateral keels anteriorly broadly flattened in section, broadly V-shaped in profile, posterior half broad, median area between them broadly rounded without a weakly angled third ridge (Fig. 44); hair of T4 pale at least laterally and often extensively, face above the antennal base mostly black → **46**

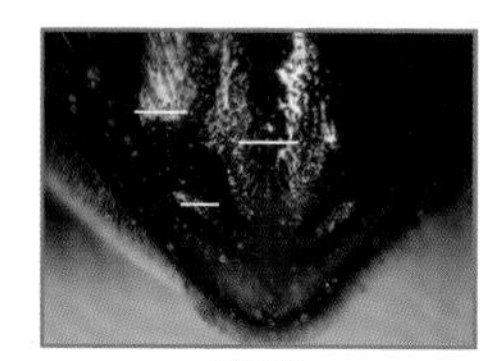

FIG 44

44a (43a) Hair of T4 yellow, face above antennal base mostly yellow → ***B. insularis***

44b Hair of T4 black, face above antennal base mostly black → **45**

45a (44b) T6 anteriorly and medially matte with large pits or punctures (Fig. 45), hair of T2–5 medium length and extensively overlapping, partly obscuring the surface; hair of the side of the thorax mostly yellow → ***B. citrinus***

FIG 45

45b T6 anteriorly and medially shiny with small pits or punctures (Fig. 46), hair of T2–5 very short, scarcely overlapping, and inconspicuous, the surface very shiny and unobscured; hair of the side of the thorax mostly black → ***B. variabilis***

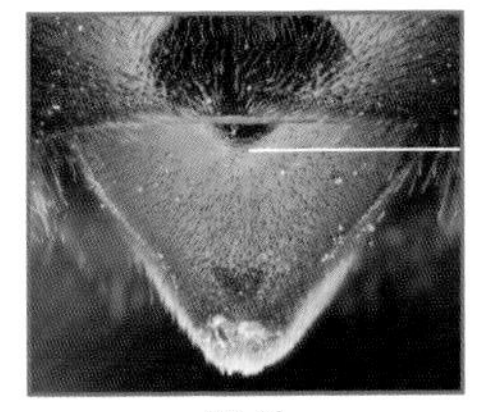

FIG 46

46a (43b) S6 with the lateral keels strong but weakly produced laterally and scarcely visible from the dorsal view projecting beyond T6 (Fig. 47); hair of the side of the thorax mostly black, T4 variable yellowish white, usually white at least posteriorly in the median area → ***B. bohemicus***

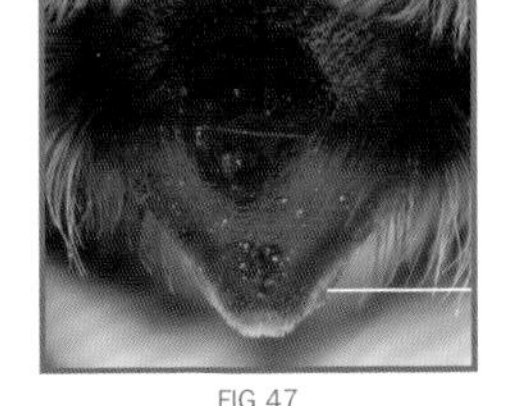

FIG 47

46b S6 with the lateral keels strongly produced laterally so clearly visible from the dorsal view projecting beyond T6 (Fig. 48); hair of the side of the thorax usually mostly yellow or at least with scattered yellow hairs in the dorsal half, T4 yellow but anteriorly with a black triangle which extends posteriorly along the midline to the posterior margin, or rarely yellow only laterally → ***B. suckleyi***

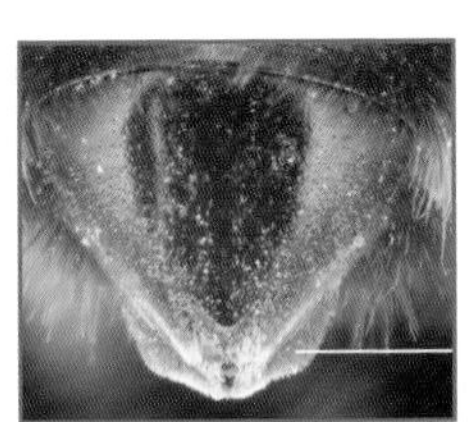

FIG 48

MALES

1a Eye similar in relative size and shape to that of any female *Bombus*, in lateral view its maximum breadth less than the maximum breadth of the area of the head posterior to the eye (gena) (Fig. 49), lateral ocellus small and located on a line between the posterior edges of the eyes, nearly twice the ocellar diameter from the eye (Fig. 50) → **3**

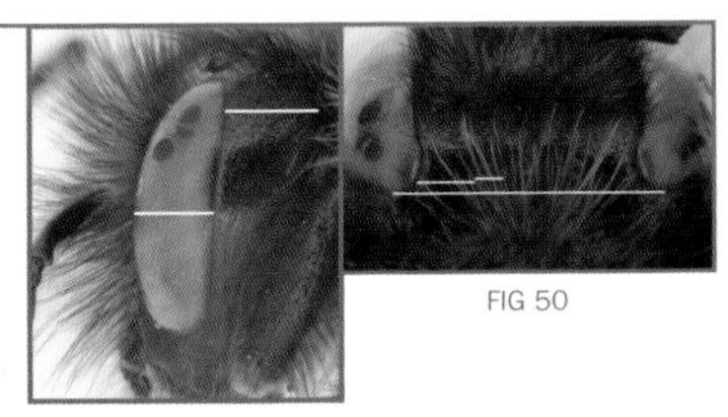

FIG 49

FIG 50

1b Eye enlarged, larger and more bulbous than that of any female *Bombus*, in lateral view its maximum breadth greater than the maximum breadth of the area of the head posterior to the eye (Fig. 51), lateral ocellus large and located anterior to a line between the posterior edges of the eyes, no more than 1× the ocellar diameter from the eye (Fig. 52) → **2**

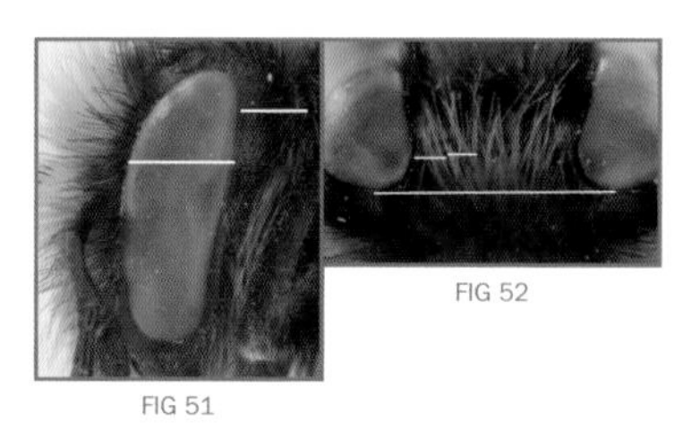

FIG 51 FIG 52

2a (1b) Eyes weakly convergent dorsally, with the distance between them dorsally >0.75× the distance between them ventrally (Fig. 53); penis-valve head dorsoventrally flattened, curved in toward the body midline and sickle-shaped (Fig. 54) → **33 (*Cullumanobombus*)**

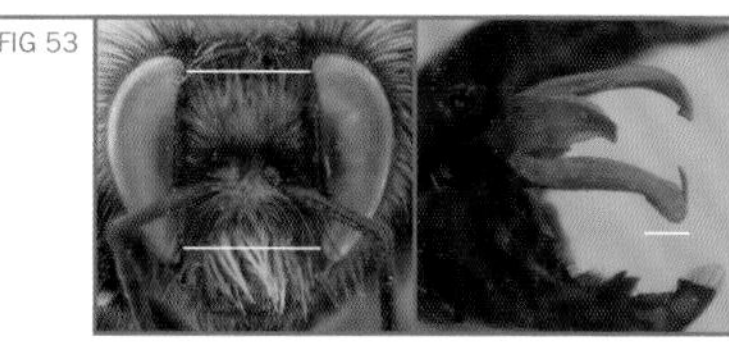

FIG 53 FIG 54

2b Eyes strongly convergent dorsally, with the distance between them dorsally half the distance between them ventrally (Fig. 55); penis-valve head laterally flattened, straight and about 5× as long as broad (Fig. 56) → **40 (*Bombias*)**

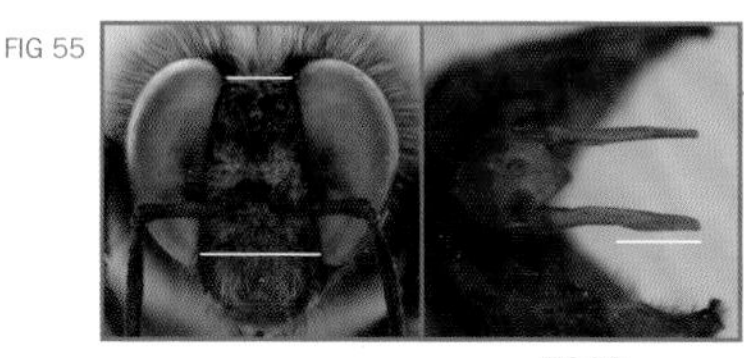

FIG 55 FIG 56

3a (1a) Antenna short, antennal flagellum (A3–13) less than 2.5× the length of the scape (A1) (Fig. 57); penis-valve head greatly broadened dorsoventrally, flared outward and forming (half of) a broad funnel shape (Fig. 58) → **29 (*Bombus*)**

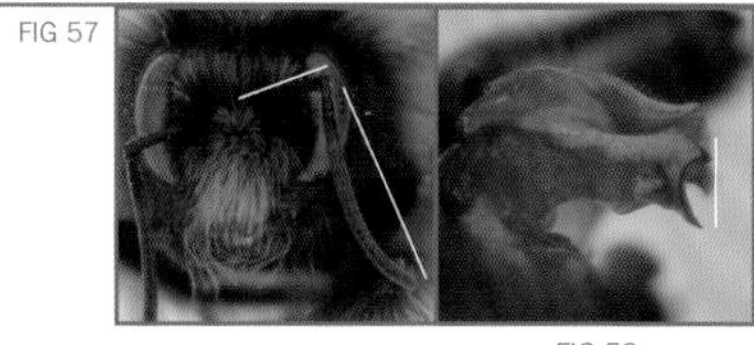

FIG 57 FIG 58

3b Antenna long or very long, antennal flagellum (A3–13) more than 2.5× the length of the scape (A1) (Fig. 59); penis-valve head either straight, or outcurved from the body midline, or incurved toward the body midline as a sickle shape, or as a short, broad, deep spoon shape (e.g., Fig. 60) → **4**

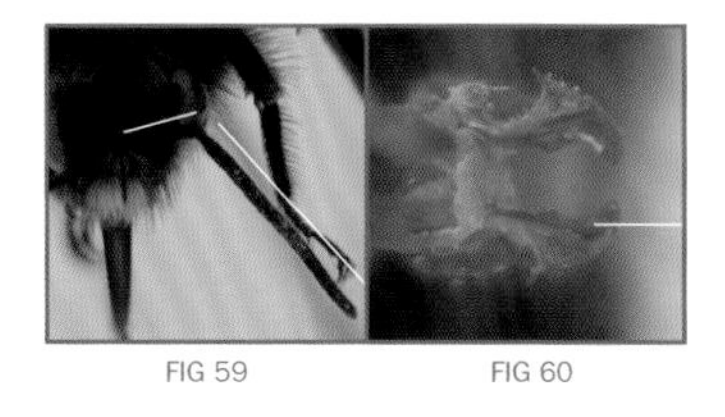

FIG 59 FIG 60

4a (3b) Volsella often yellow (weakly sclerotized), without distal (posterior) hooks on the inner (medial) edge (Fig. 61), gonostylus inner process with many long branched hairs (Fig. 61) → **41 (*Psithyrus*)**

FIG 61

4b Volsella medium to dark brown (strongly sclerotized), with at least one short distal hook on the inner (medial) edge (Fig. 62), gonostylus inner process without long branched hairs (Fig. 62) → **5**

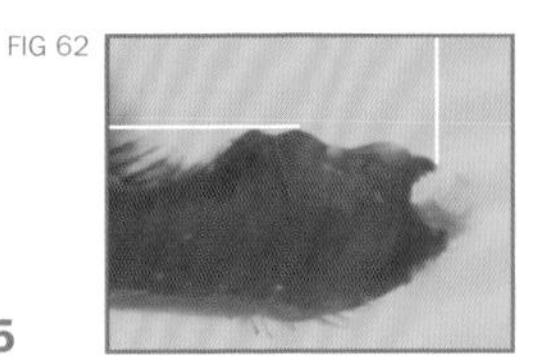

FIG 62

5a (4b) Antenna long, antennal flagellum (A3–13) 2.5–3× the length of the scape (A1) (Fig. 63); penis-valve head strongly incurved toward the body midline in a flat sickle shape (Fig. 64), gonostylus inner (medial) margin usually with at least a trace of a submarginal parallel groove along much of its length that ends anteriorly and posteriorly without produced corners (Fig. 65) → **8 (*Pyrobombus*)**

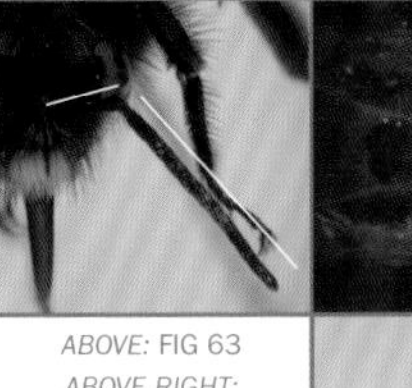

ABOVE: FIG 63
ABOVE RIGHT: FIG 64

FIG 65

5b Antenna very long, antennal flagellum (A3–13) more than 3× the length of the scape (A1) (Fig. 66); penis-valve head either straight, or outcurved from the body midline (Fig. 67), or incurved toward the body midline as a short, broad, deep spoon shape, gonostylus inner (medial) margin without a submarginal parallel groove along much of its length, *or if* the groove is present *then* it ends anteriorly and posteriorly with produced corners (Fig. 68) → **6**

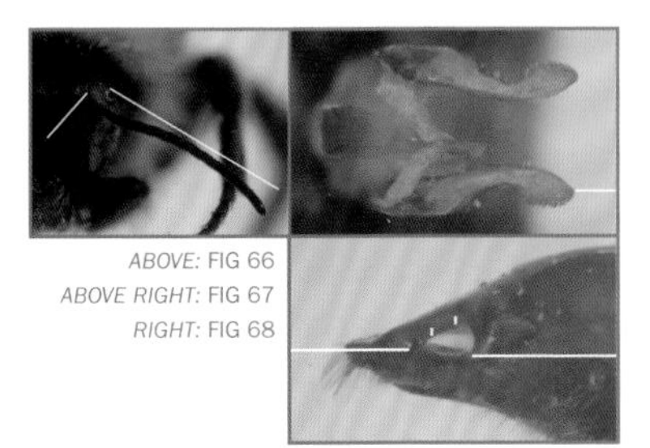

ABOVE: FIG 66
ABOVE RIGHT: FIG 67
RIGHT: FIG 68

6a (5b) Penis-valve head incurved strongly toward the body midline as a short, broad, deep spoon shape (Fig. 69) → **38 (*Subterraneobombus*)**

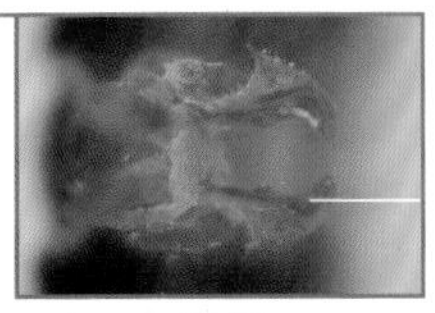

FIG 69

6b Penis-valve head either straight, or outcurved slightly from the body midline (Fig. 70) → **7**

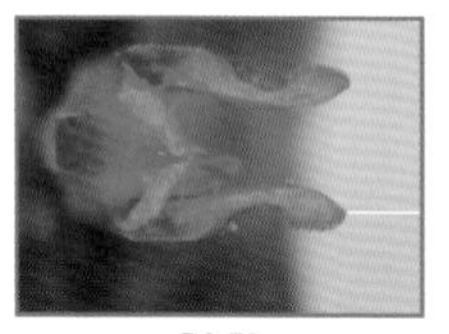

FIG 70

7a (6b) Hindleg tibia posterior fringe hairs longer than the maximum breadth of the tibia (Fig. 71), hindleg tibia outer surface center smooth and shining, without short hairs (Fig. 71); penis-valve head straight, weakly dorsoventrally flattened, its outer edge with 2–3 teeth (Fig. 72), gonostylus without a strong proximal projection in toward the body midline, inner (medial) margin usually with a submarginal parallel groove along much of its length, often with spines at the ends (Fig. 73) → **26 (*Alpinobombus*)**

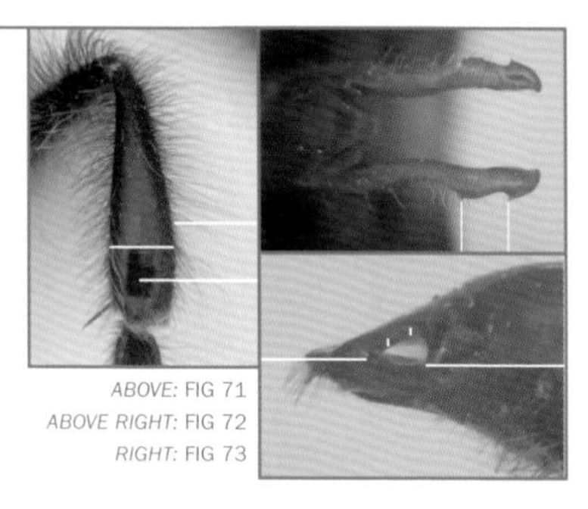

ABOVE: FIG 71
ABOVE RIGHT: FIG 72
RIGHT: FIG 73

7b Hindleg tibia posterior fringe hairs shorter than the maximum breadth of the tibia (Fig. 74), hindleg tibia outer surface center rough and matte, with many short hairs (Fig. 74); penis-valve head outcurved slightly from the body midline, its outer edge with many teeth (Fig. 75), gonostylus with a strong flattened proximal projection in toward the body midline, inner (medial) margin without a submarginal parallel groove (Fig. 76) → **37 (*Thoracobombus*)**

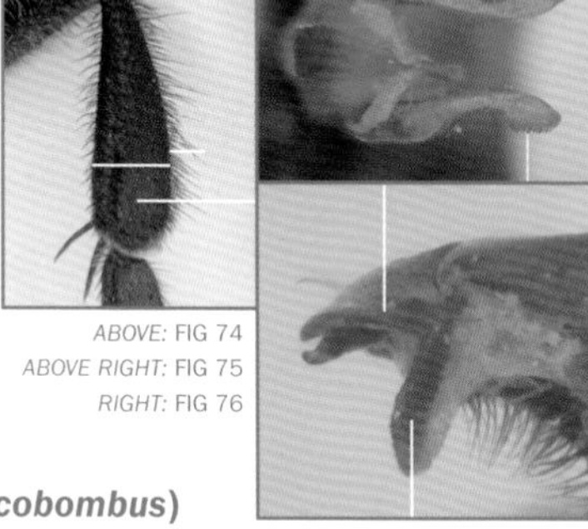

ABOVE: FIG 74
ABOVE RIGHT: FIG 75
RIGHT: FIG 76

8a (5a) *Pyrobombus*
Penis-valve head with the distal recurved part of the "sickle" long and narrow, at least 3× longer than the breadth of the distal half and usually less than half the breadth of the neighboring straight proximal part of the penis-valve head (Fig. 77), gonostylus inner (medial) edge usually slightly concave (Fig. 78) → **9**

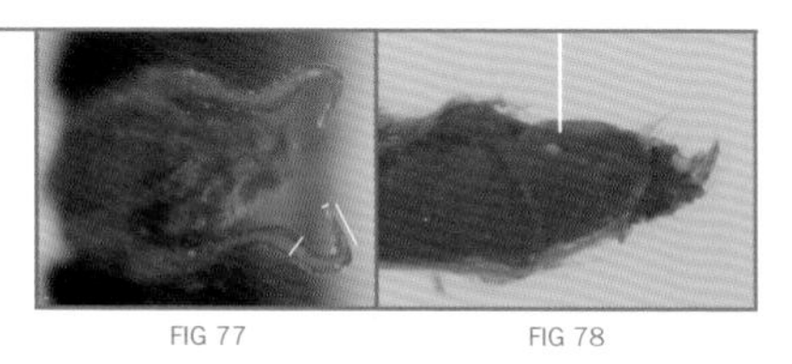

FIG 77 FIG 78

8b Penis-valve head with the distal recurved part of the "sickle" short and broad, little more than 2× longer than the breadth of the distal half and similar in breadth to the neighboring straight proximal part of the penis-valve head (Fig. 79), gonostylus inner (medial) edge usually nearly straight or convex (Fig. 80) → **16**

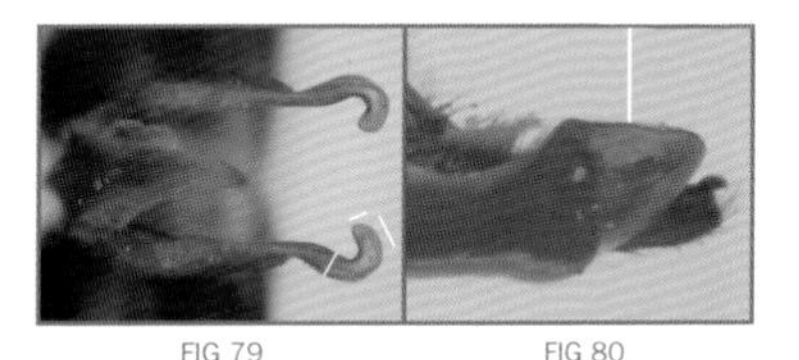

FIG 79 FIG 80

9a (8a) Penis-valve shaft with the strong ventrolateral angle located at about 0.25× the distance from its base to the proximal end of the distal head (Fig. 81), the penis-valve head with the distal recurved "sickle" often very narrow but ending distally in a slight expansion that is rounded (Fig. 82) → ***B. sylvicola***

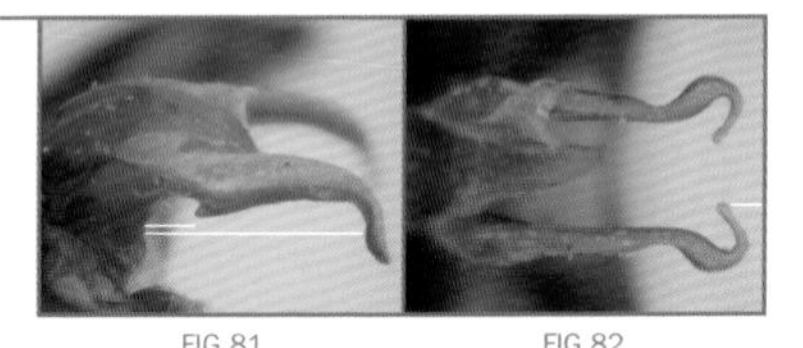

FIG 81 FIG 82

9b Penis-valve shaft with the strong ventrolateral angle located at about 0.33× the distance from its base to the proximal end of the distal head (Fig. 83), the penis-valve head with the distal recurved "sickle" narrow and ending distally by further narrowing to a point, or at most rounded without being expanded (Fig. 84) → **10**

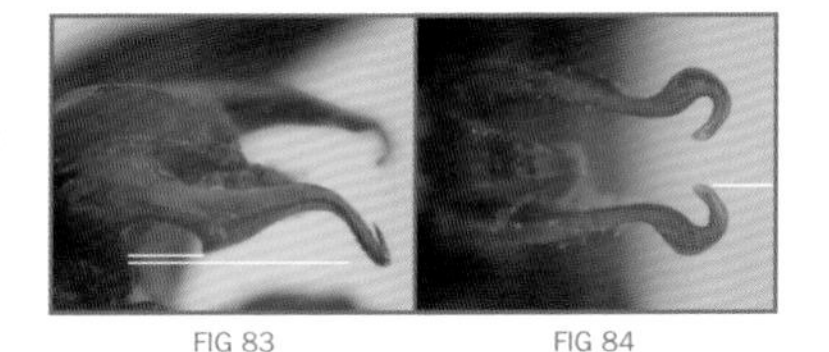

FIG 83 FIG 84

10a (9b) Penis-valve head with the distal recurved "sickle" long and very narrow, scarcely flattened and about 2× broader than thick, almost spinelike (Fig. 85) → **12**

FIG 85

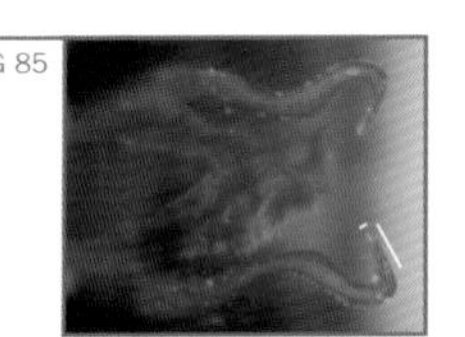

10b Penis-valve head with the distal recurved "sickle" long and narrow, but distinctly flattened and more than 3× broader than thick (Fig. 86) → **11**

FIG 86

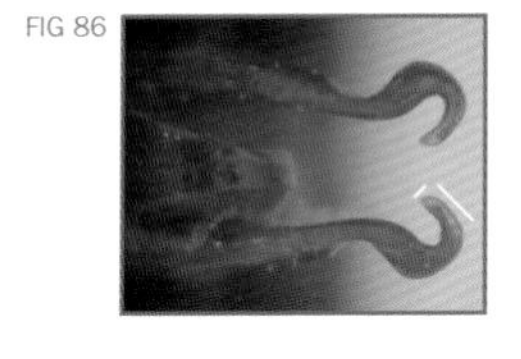

11a (10b) Hair of the thorax predominantly bright yellow, at most the thoracic dorsum between the wings with a broad black spot → ***B. bimaculatus***

11b Hair of the thorax yellow, but at least all the thoracic dorsum anterior to and between the wings with many black hairs densely intermixed, often with a diffuse black band between the wings → ***B. melanopygus***

12a (10a) Hair of T2–4 entirely black → ***B. impatiens***

12b Hair of at least some of T2–4 yellow or orange → **13**

13a (12b) Hair of the thorax predominantly black, the thoracic dorsum anterior to the wings with a yellow band, occasionally on the dorsum elsewhere with yellow intermixed, T1 black → ***B. vosnesenskii***

13b Hair of the thorax predominantly yellow, dorsally between the wings with a black band, T1 yellow → **14**

14a (13b) Hair of the thoracic dorsum anterior and posterior to the wings (scutellum) yellow without black hairs → ***B. huntii***

14b Hair of the thoracic dorsum anterior to the wings yellow often with at least a few black hairs intermixed, posterior to the wings (scutellum) yellow but anteriorly and medially with at least a few black hairs → **15**

15a (14b) Hair of T2 entirely orange → ***B. ternarius***

15b Hair of T2 anteriorly and medially extensively yellow, laterally with at least some black → ***B. bifarius***

16a (8b) Penis-valve head with the distal recurved "sickle" broad, flat, and triangular with the sides converging consistently to a rounded point (Fig. 87), gonostylus proximally on its inner (medial) edge usually with a very deep notch as broad and half as deep as the penis-valve head breadth, separating the straight inner (medial) edge from the gonocoxa (Fig. 88) → ***B. perplexus***

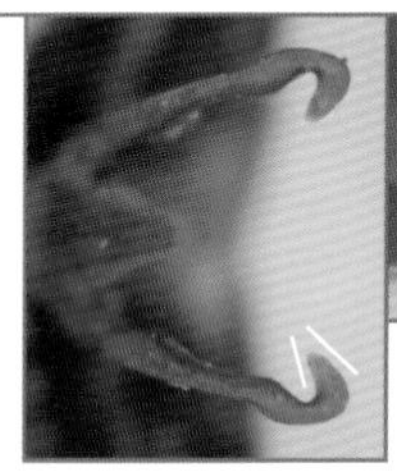

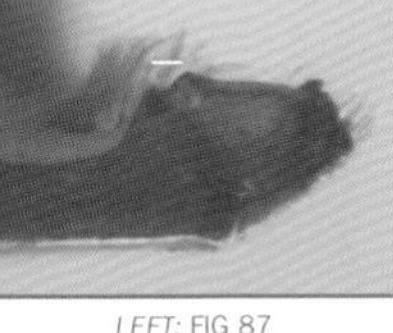

LEFT: FIG 87
ABOVE: FIG 88

16b Penis-valve head with the distal recurved "sickle" broad and flat but slightly hammer- or club-shaped, with the sides parallel or diverging to a bluntly rounded end (Fig. 89), gonostylus proximally on its inner (medial) edge usually with at most only a very shallow notch separating the inner (medial) edge from the gonocoxa, the notch usually much narrower than the penis-valve head (Fig. 90) → **17**

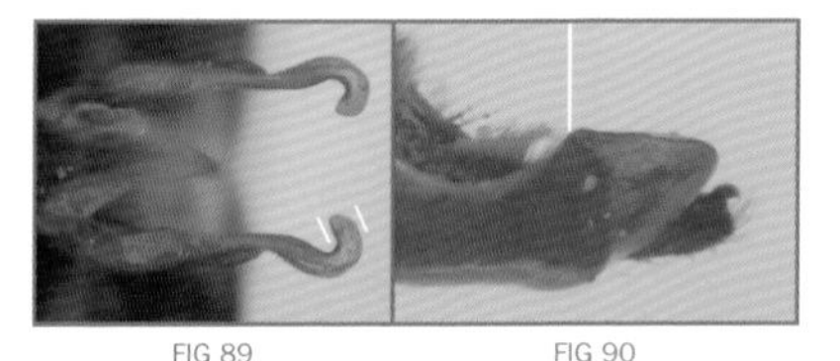

FIG 89 FIG 90

17a (16b) Antennal segment A3 long, length nearly twice its maximum breadth, almost as long as antennal segment A5 (Fig. 91); gonostylus inner (medial) margin thin with at most only a short very indistinct parallel submarginal groove (Fig. 92) → **18**

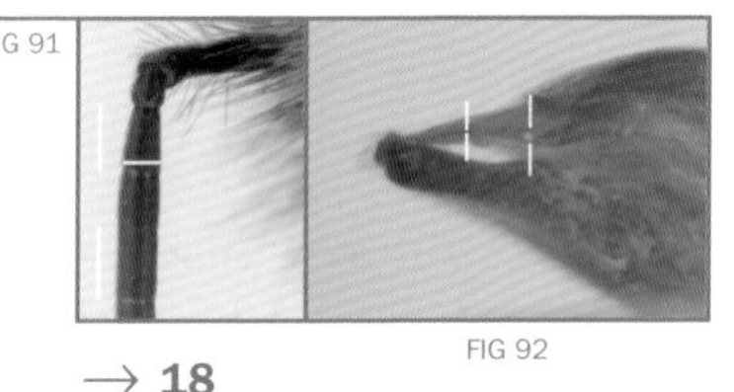

FIG 91 FIG 92

17b Antennal segment A3 short, length less than 1.5× maximum breadth, much shorter than antennal segment A5 (Fig. 93); gonostylus inner (medial) margin thick with a distinct parallel submarginal groove (Fig. 94) → **22**

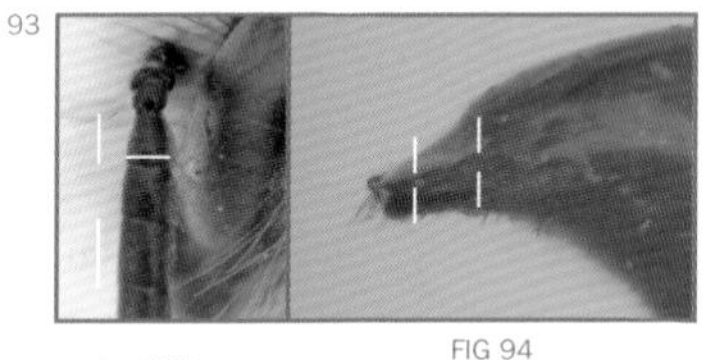

FIG 93 FIG 94

18a (17a) Hair of T1–2 black → ***B. caliginosus***

18b Hair of T1–2 extensively yellow → **19**

19a (18b) Hair of the face yellow with black intermixed at least around the antennal base, thoracic dorsum usually without a black band between the wings and often almost entirely yellow, T3 usually black at least medially and posteriorly, occasionally with yellow intermixed medially → ***B. vagans***

19b Hair of the face yellow with black hairs mostly confined to the edges, thoracic dorsum usually with a band of black at least intermixed between the wings, T3 usually yellow or orange-red, but *if* black *then* with a yellow posterior fringe → **20**

20a (19b) Hair usually predominantly yellow, T3–4 usually yellow without black intermixed, but rare dark males have the yellow only anteriorly on the thoracic dorsum and on T2–3, with black on the thorax posterior to the wings and on T1 → ***B. vandykei***

20b Hair of T3–4 either orange-red, or *if* yellow *then* with black intermixed anteriorly → **21**

21a (20b) Hair of the thoracic dorsum anteriorly yellow with few or no black hairs intermixed near the midline so that the black band between the wings is distinct, or sometimes the anterior edge of the black band is indistinct because of black hairs anteriorly to the black band, the thoracic dorsum posterior to the wings yellow but often with many scattered black hairs especially near the midline, T3–4 either red or extensively yellow, with at least a few black hairs intermixed near the anterior lateral corners or sometimes forming broad anterior black bands → ***B. flavifrons***

21b Hair of the thoracic dorsum anteriorly yellow with no black intermixed near the midline so that the black band between the wings is sharply distinct, or sometimes the anterior edge of the black band is indistinct because there are more yellow hairs intermixed posteriorly within the black band and especially laterally, the thoracic dorsum posterior to the wings yellow, usually without black hairs except for a few anteriorly near the midline, T3–4 red without any black hairs → ***B. centralis***

22a (17b) Antennal segments A3–4 posterior edges each with a patch of dense short hair and/or bristles (Fig. 95) → **23**

FIG 95

22b Antennal segments A3–4 posterior edges without dense patches of short hair and bristles, at most with just one or two short bristles (Fig. 96) → **24**

FIG 96

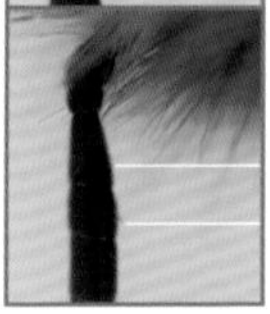

23a (22a) Antennal segments A3–8 not broadened or thickened, antennal segments A3–4 posterior edges each with a patch of dense short hair and bristles, segments A4–12 without similar obvious patches (Fig. 97); hair of thoracic dorsum anteriorly yellow, T3–7 black, T5 especially sometimes with some yellow or brownish → ***B. sandersoni***

FIG 97

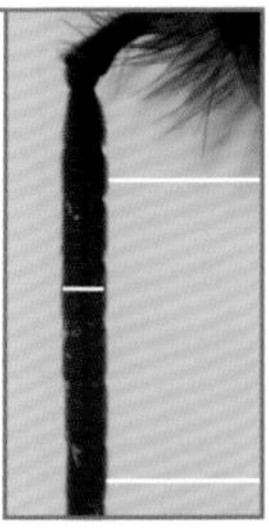

23b Antennal segments A3–8 slightly but distinctly broadened or thickened, antennal segments A3–4 posterior edges each with a patch of dense medium-length hair, segments A4–12 with similar patches of less obvious short hair (Fig. 98); hair of thoracic dorsum anteriorly yellow usually with black intermixed, T3 black or posteriorly orange, T4–7 usually extensively pale orange but occasionally black → ***B. mixtus***

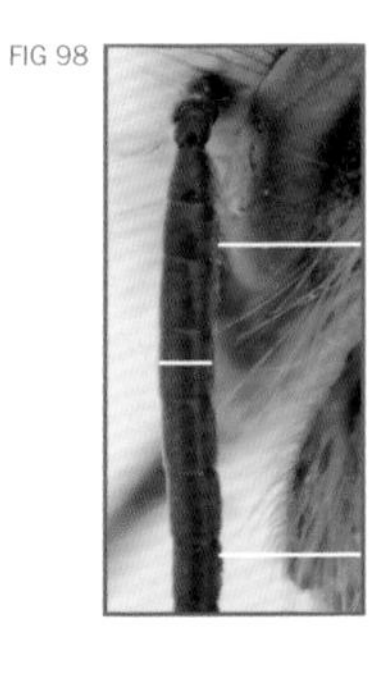
FIG 98

24a (22b) Hair of the thoracic dorsum with a weakly defined black band between the wings, posterior to the wings with yellow and black intermixed, T3 anteriorly black and posteriorly broadly yellow → ***B. sitkensis***

24b Hair of the thoracic dorsum with a clearly defined black band between the wings, posterior to the wings yellow, T3 black, at most with a very weak, pale fringe → **25**

25a (24b) Hair of T6–7 orange, which is sometimes pale → ***B. frigidus***

25b Hair of T6–7 white → ***B. jonellus***

26a (7a) *Alpinobombus*
Hair of the face and T5–7 almost entirely black, T5–7 very rarely with some orange hairs intermixed; penis-valve head on its outer edge proximal to the apex by a distance approximately equal to the breadth of the penis-valve head without a tooth, rarely with a very small point without breaking the longer curve of the outer penis-valve head (Fig. 99) → ***B. hyperboreus***

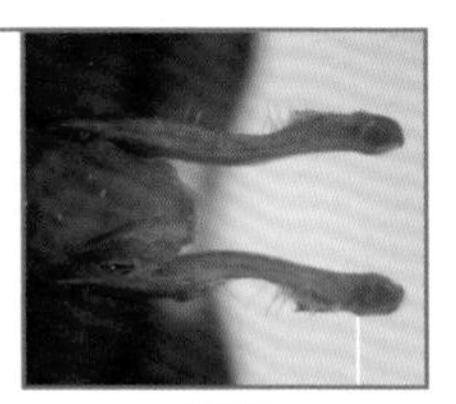

FIG 99

26b Hair of the face usually with a patch of yellow, or *if* the face is black *then* T5–7 with many orange hairs; penis-valve head on its outer edge proximal to the apex by a distance approximately equal to the breadth of the penis-valve head with a tooth placed on a convex bump on the edge of the penis-valve head, although this tooth may be small (Fig. 100) → **27**

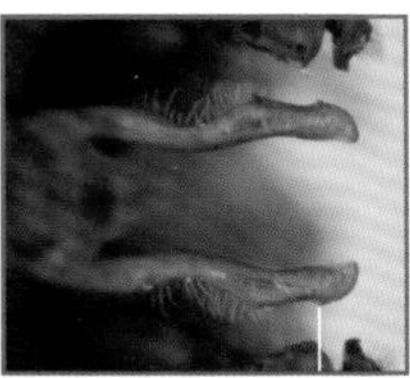

FIG 100

27a (26b) Hair of the side of the thorax yellow, posteriorly with more or less discrete patches of black, thoracic dorsum between the wings black with few yellow hairs intermixed, T3 often predominantly yellow, or *if* T3 black *then* T2 is also predominantly black, T4–5 predominantly black with a small amount of yellow laterally on T4 and orange posteriorly on T5; penis-valve head on its outer edge proximal to the apex by a distance approximately equal to the breadth of the penis-valve head with a small weak tooth, margin between the outer teeth straight (Fig. 101) → ***B. neoboreus***

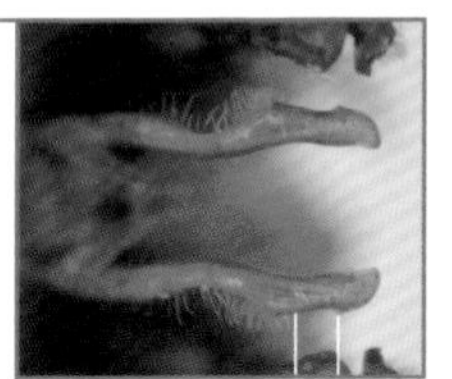

FIG 101

27b Hair of the side of the thorax yellow, with or without black intermixed, thoracic dorsum between the wings black with many yellow hairs intermixed, or sometimes the thorax predominantly black, T3 may be black, or *if* T3 with yellow *then* T4–5 usually also with extensive yellow or orange; penis-valve head on its outer edge proximal to the apex by a distance approximately equal to the breadth of the penis-valve head with a large strong tooth, margin between the outer teeth convex (Fig. 102) → **28**

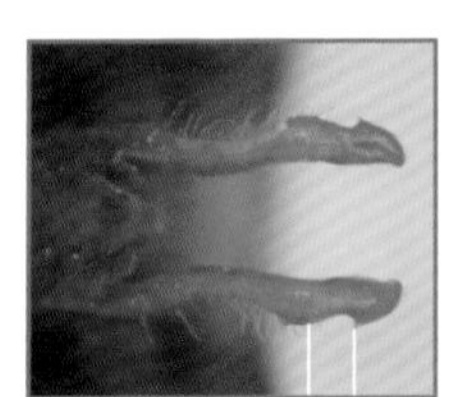

FIG 102

28a (27b) Hair of the thorax predominantly yellow, T3–4 black with yellow laterally, often as a lateral longitudinal fringe that continues around as strongly differentiated posterior fringes, or T4 sometimes orange, S2–4 entirely yellow; gonostylus inner (medial) edge strongly and consistently concave, the submarginal groove broad and sharply defined, the anterior and posterior corners strongly extended (Fig. 103) → ***B. balteatus***

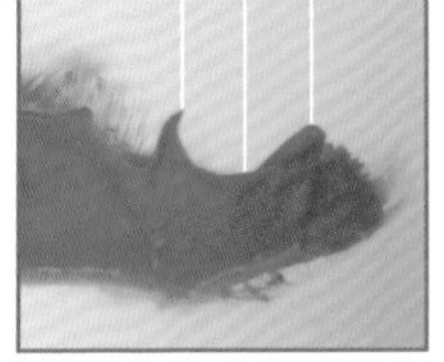

FIG 103

28b Hair of the thorax sometimes predominantly black, or the side of the thorax and T3–4 yellow often with black extensively intermixed, T3–4 posterior fringes weakly differentiated, S2–4 with yellow and black intermixed, or black; gonostylus inner (medial) edge weakly concave with a convexity near its midpoint, the submarginal groove indistinct, the anterior and posterior corners weakly rounded (Fig. 104) → ***B. polaris***

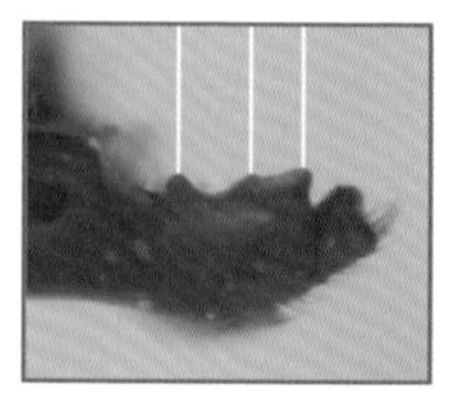

FIG 104

29a *Bombus s. str.*
(3a) Hair of T2 (when viewed from the side) at least anteriorly more distinctly brown than the yellow on the rest of the body, T3–7 entirely black → ***B. affinis***

29b Hair of T2 yellow or black, T3–7 with some brown, yellow, or white → **30**

30a (29b) Hair of the thoracic dorsum between the wings extensively yellow, T1–4 black → ***B. franklini***

30b Hair of the thoracic dorsum between the wings extensively black, T1–4 with some pale hair at least posteriorly on T4 → **31**

31a (30b) Hair of T2 yellow, T3 black or with the posterior fringe yellow, T5–7 white → ***B. cryptarum***

31b Hair of T2 extensively black, or *if* T2 yellow *then* T3 also yellow *and* T5–7 extensively black → **32**

32a (31b) Hair of T2–3 entirely yellow, T4–6 extensively black but often with some yellow-orange on T6 → ***B. terricola***

32b Hair of T2 yellow with black in the middle anteriorly and T3 yellow, or T2–3 black throughout, T4 posteriorly and T5–6 extensively white or yellow-orange → ***B. occidentalis***

33a (2a) *Cullumanobombus*
Lateral ocellus about one ocellar diameter from the eye (Fig. 105); gonostylus with the inner (medial) edge with two deep concavities (Fig. 106), volsella projecting beyond the distal end of the gonostylus by more than the volsellar breadth at that point (Fig. 106) → ***B. rufocinctus***

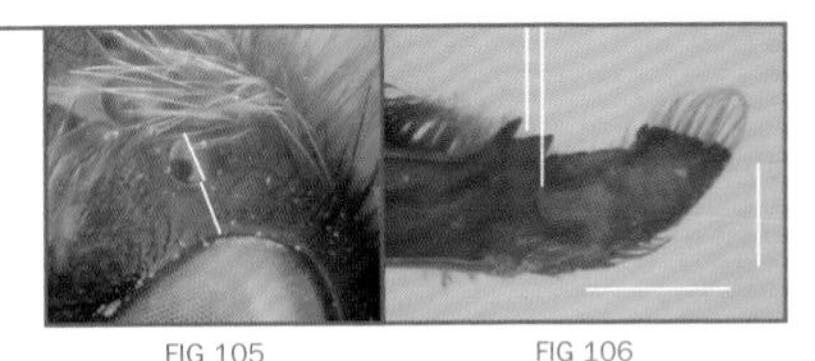
FIG 105 FIG 106

33b Lateral ocellus less than half an ocellar diameter from the eye (Fig. 107); gonostylus with the inner (medial) edge with one deep concavity (Fig. 108), volsella projecting beyond the distal end of the gonostylus by less than the volsellar breadth at that point (Fig. 108) → **34**

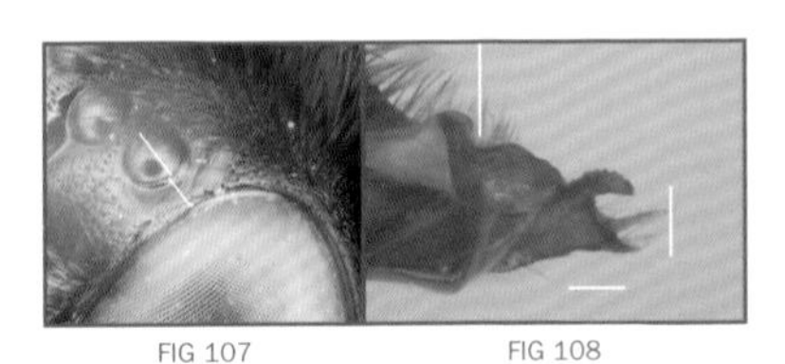
FIG 107 FIG 108

34a (33b) Mandible with the beard on the posterior margin shorter than the maximum breadth of the mandible (Fig. 109); penis-valve head with the outer flange extended proximally as a hook (Fig. 110); hair of T3 lying completely flat on the body surface → ***B. fraternus***

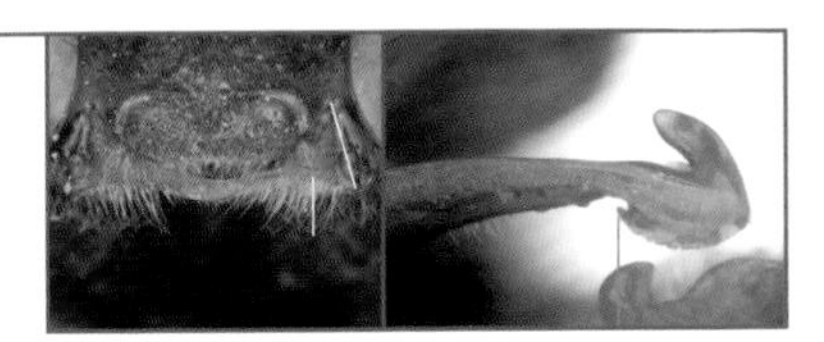
LEFT: FIG 109
ABOVE: FIG 110

34b Mandible with the beard on the posterior margin longer than the maximum breadth of the mandible (Fig. 111); penis-valve head with the outer flange without a proximal hook (Fig. 112); hair of T3 at least semi-erect from the body surface → **35**

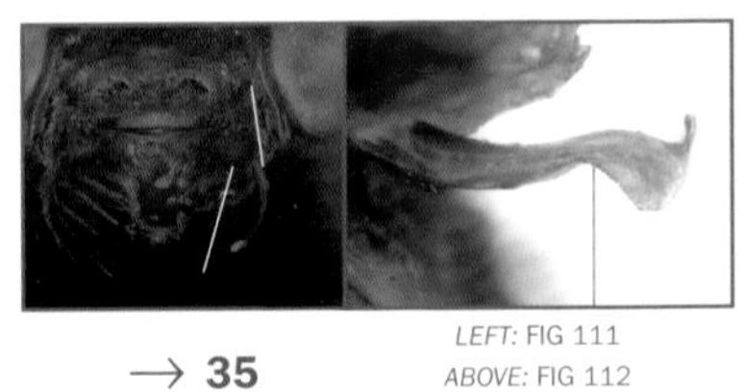
LEFT: FIG 111
ABOVE: FIG 112

35a (34b) Hair of the thoracic dorsum yellow, at most with only a few short, inconspicuous black hairs, T3 entirely yellow → ***B. morrisoni***

35b Hair of the thoracic dorsum yellow between the wings, usually with a broad black spot or broad band of black, T3 entirely black or with small patches of yellow laterally on the posterior margin → **36**

36a (35b) Hair of the side of the thorax entirely yellow, thoracic dorsum between the wings usually with a broad black spot; penis valve with the recurved head broad, no longer than broad, as an equilateral triangle (Fig. 113) → ***B. griseocollis***

FIG 113

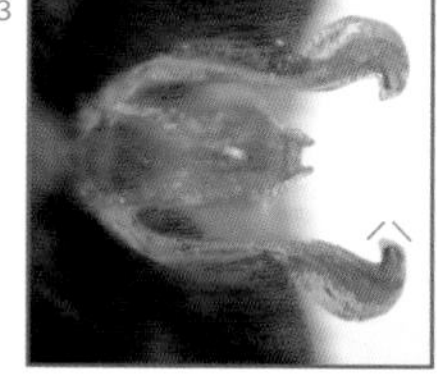

36b Hair of the side of the thorax extensively black, thoracic dorsum between the wings with a broad black band; penis valve with the recurved head narrow, much longer than broad (Fig. 114) → ***B. crotchii***

FIG 114

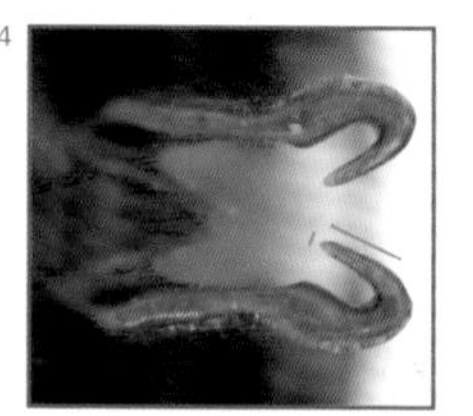

37a (7b) *Thoracobombus*

Hair of T7 black, or *if* T7 orange *then* T2–3 extensively black; penis-valve head nearly as broad as long and triangular (Fig. 115) → ***B. fervidus***

FIG 115

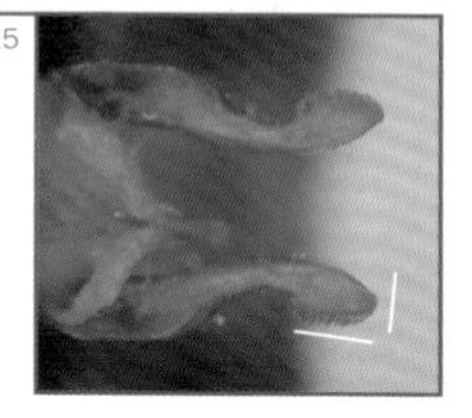

37b Hair of T7 often orange, or *if* T7 black *then* T2–3 entirely yellow; penis-valve head more than twice as long as broad and banana-shaped (Fig. 116) → ***B. pensylvanicus***

FIG 116

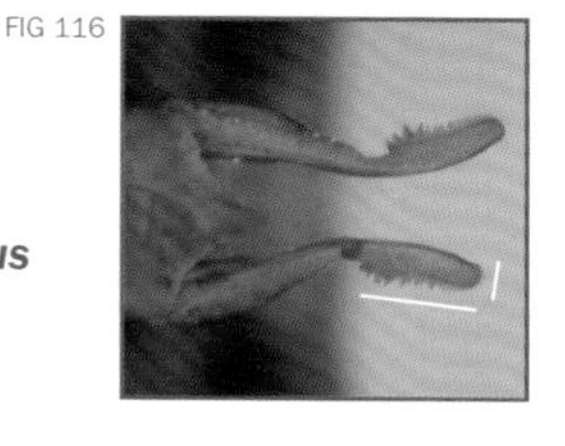

38a *Subterraneobombus*
(6a) [Male not yet seen from N America]
Penis valve with the ventrolateral process trident-shaped but the dorsal tooth very small, the middle tooth broadly rounded and only weakly marked (Fig. 69) → ***B. distinguendus***

FIG 117

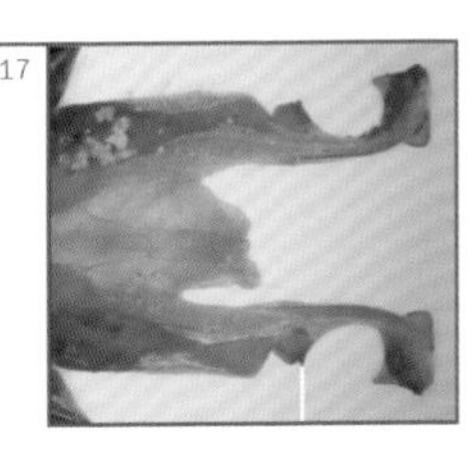

38b Penis valve with the ventrolateral process trident-shaped with the dorsal tooth strongly produced as a narrow spine, the middle tooth produced as a broad equilateral triangle (Fig. 70) → **39**

FIG 118

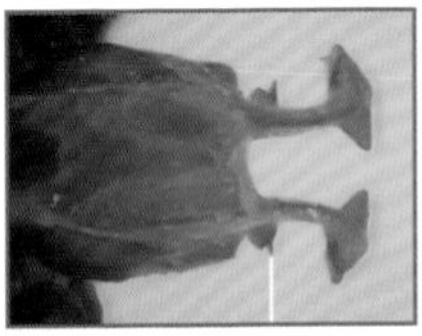

39a Hair of the head and the anterior thoracic
(38b) dorsum similar shades of gray-white or yellow, side of the thorax with only a few black hairs intermixed posteriorly, T1–5 orange-brown or yellow, T5 without black → ***B. appositus***

39b Hair of the face between the clypeus and the ocelli pale yellow, hair of the head dorsally and the anterior thoracic dorsum a distinctly darker yellow, side of the thorax with many black hairs intermixed almost throughout, T1–4 an even darker golden yellow, T5 with at least an anterior black band → ***B. borealis***

40a *Bombias*
(2b) Hair of T6–7 black; area anterior to the ocelli with a band of large pits or punctures with the spaces between them rough and dull with many small punctures (Fig. 119) → ***B. auricomus***

FIG 119

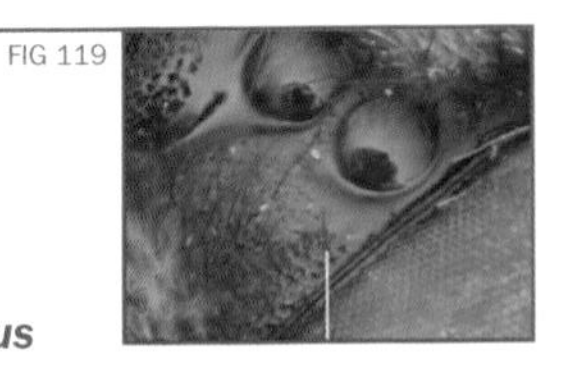

40b Hair of T6–7 extensively orange; area anterior to the ocelli with few large pits or punctures and with the spaces between them with few small punctures, smooth and shining (Fig. 120) → ***B. nevadensis***

FIG 120

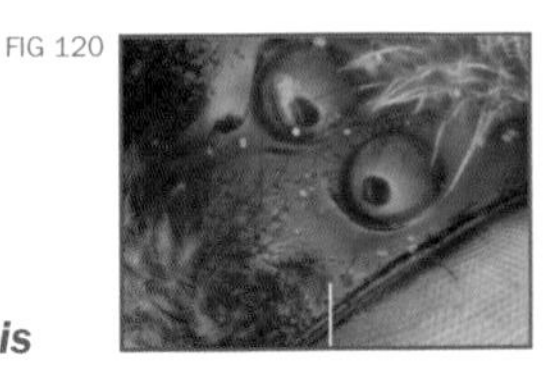

41a (4a) *Psithyrus*
Hair of T6–7 orange in part, often with paler tips; volsella narrow, sides parallel or diverging, and finger-shaped (Fig. 121) → ***B. flavidus***

FIG 121

41b Hair of T6–7 black or black and yellow; volsella broad, sides converging, and triangular (Fig. 122) → **42**

FIG 122

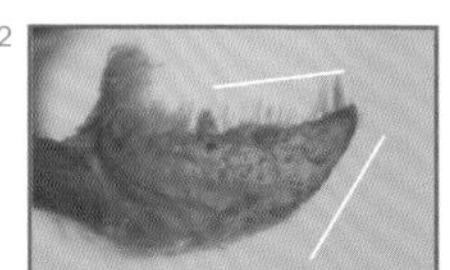

42a (41b) Penis valve with the ventrolateral angle much reduced and not visible from the dorsal view (Fig. 123) → **43**

FIG 123

42b Penis valve with the ventrolateral angle broadly triangular and visible from the dorsal view (Fig. 124) → **44**

FIG 124

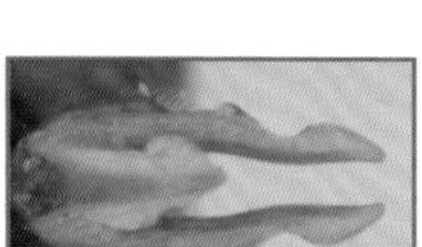

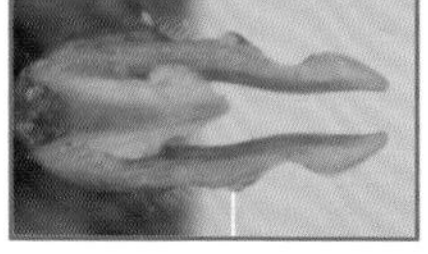

43a (42a) Hair of T2 entirely black, side of the thorax extensively black; penis-valve head about 2× as long as broad (Fig. 125) → ***B. bohemicus***

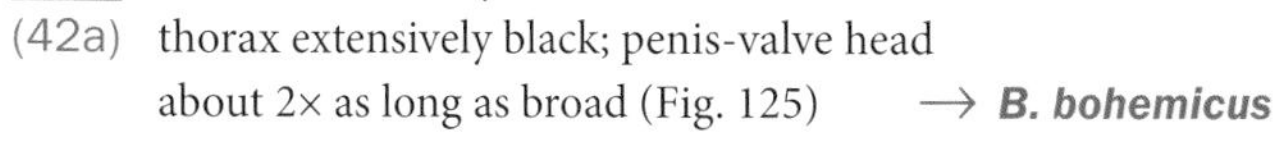

FIG 125

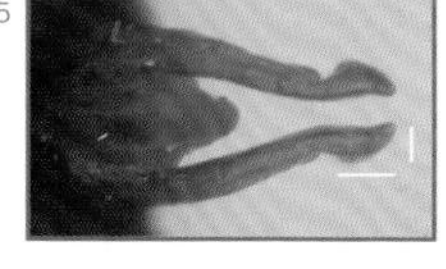

43b Hair of T2 extensively yellow, side of the thorax at most with black hairs posteriorly; penis-valve head nearly 3× as long as broad (Fig. 126) → ***B. suckleyi***

FIG 126

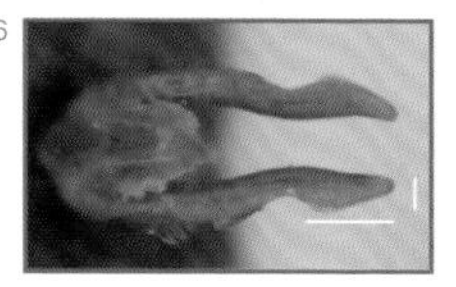

44a (42b) Hair of T4–7 entirely black → ***B. citrinus***

44b Hair of T4–7 with some yellow, sometimes T4 only dark brownish yellow laterally → **45**

45a (44b) Hair of the face with many yellow hairs above the base of the antenna, hair of T3 anteriorly and medially as long as or longer than the hindleg basitarsus breadth; volsella maximum breadth in dorsal perpendicular view about 1.5× that of the narrowest part of the distal gonocoxa (Fig. 79) → ***B. insularis***

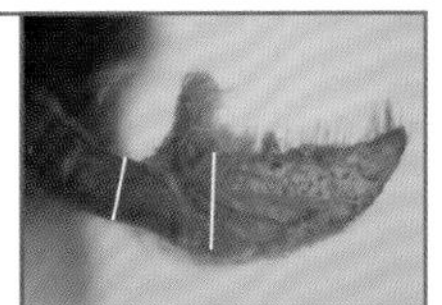

FIG 127

45b Hair of the face black, hair of T3 anteriorly and medially shorter than the hindleg basitarsus breadth; volsella maximum breadth in dorsal perpendicular view nearly twice that of the narrowest part of the distal gonocoxa (Fig. 80) → ***B. variabilis***

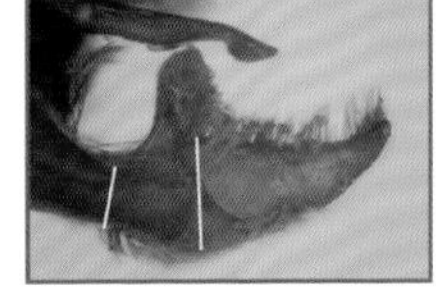

FIG 128

GLOSSARY

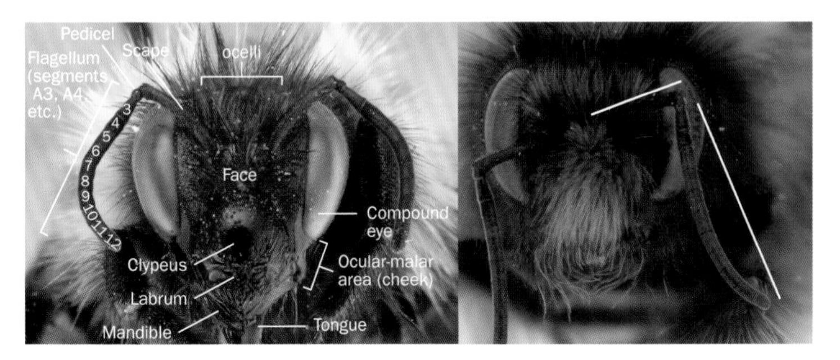

Head morphology.

Antenna.

Antenna (pl. antennae) One of a pair of simple, elbowed, sensory "feelers" attached to the face of a bumble bee and consisting of three segments: scape, pedicel, and flagellum.

Antennal base Where the antenna joins the head (see *antenna*).

Antennal segments (A#) The articles of the antenna; morphologically there are only three antennal segments as defined by muscle attachment: the scape, pedicel and flagellum. The flagellum is subdivided into units that are often called "segments" but should be called flagellomeres (see also *antenna*, *flagellum*).

Anterior Toward the front.

Apex The tip of a structure, at the opposite end from the base.

Basitarsus The first element of the tarsus, much enlarged in bumble bees; sometimes called the metatarsus (see *tarsus*).

Breadth Distance between the sides of a structure.

Leg morphology.

Cheek (oculo-malar area or malar space) The area between the bottom of the compound eye and the base of the mandible. Sometimes simply called the malar area, it is the lower part of the gena (see *gena*).

Clypeus The area below the antennae, bounded above and laterally by the frontoclypeal (epistomal) suture and to which the labrum is attached below.

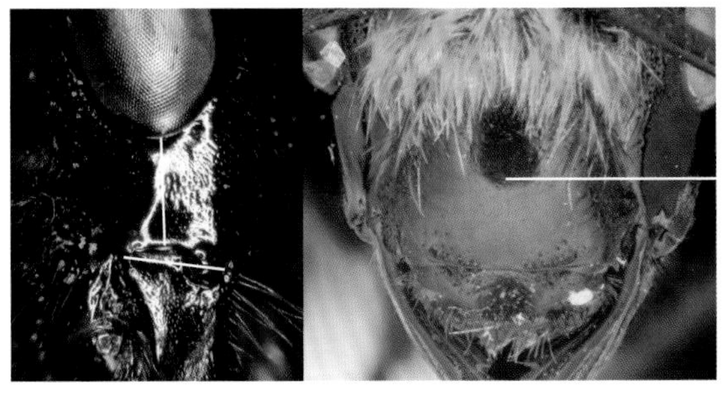

Cheek (ocular-malar area).

Clypeus.

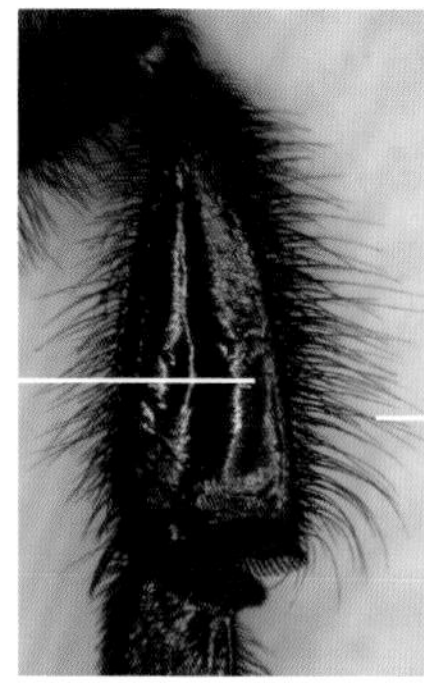
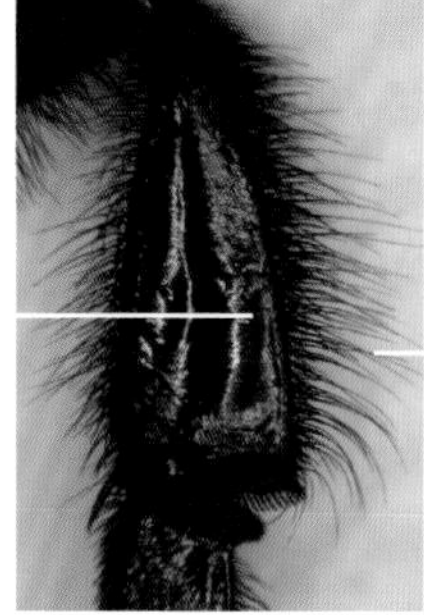

Corbicula.

Concave A surface bending inward.
Concavities Depressions in a surface.
Convergent Coming together.
Convex A surface bending outward.
Corbicula (pl. corbiculae) A bare nearly flat area surrounded by long hairs, used by female bees to transport pollen. In bumble bees it is on the outer surface of the expanded tibia of the female hindleg.
Exoskeleton The rigid external skeleton, which in a bee is largely composed of chitin; it has flexible areas called "joints" to allow for movement.
Distal Away from the body, toward the free end of a structure.
Dorsal Upperside, above, or on top.
Flagellomeres Articulated subdivisions (units) of the flagellum sometimes referred to as "segments", but lacking muscle attachments of true segments, numbering 10 in female bumble bees or 11 in males (see *antenna*).
Flagellum The third true segment of the antenna, subdivided into units called flagellomeres (see *antenna*).
Gena The area of the head behind and below the compound eye (see *cheek*).
Gonocoxa (pl. gonocoxae) Paired outer lateral arms of the male genital capsule, tipped with the gonostylus, used to clasp the female during mating (see *gonostylus*).

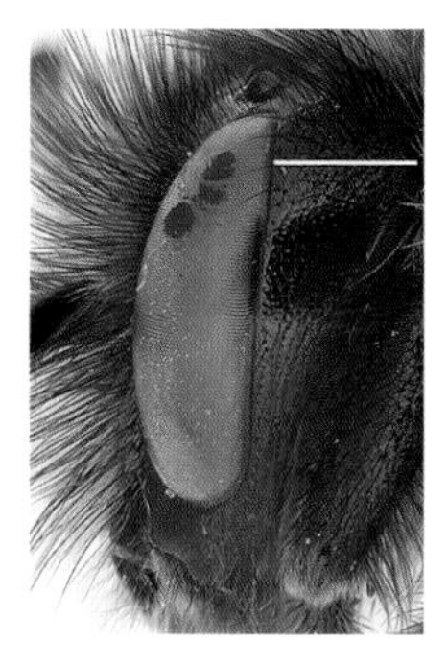

Gena.

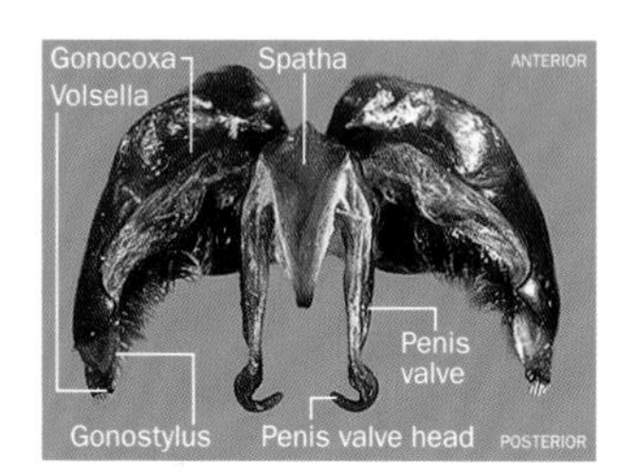

Male genitalia morphology.

Gonostylus (pl. gonostyli) The small plate at the apex of the gonocoxa, forming part of the male claspers, used to clasp the female during mating.
Groove An elongate depression on the surface of the exoskeleton.
Hindleg The leg of the third mesosomal (thoracic) segment.
Inner eye margin Margin adjacent to the compound eye closest to the ocelli on the upperside of the head.
Keel An elevated ridge (as the lateral keels on S6 in cuckoo bumble bee females).

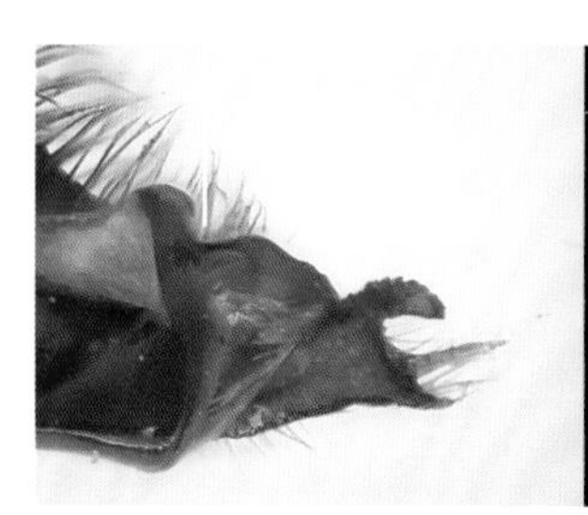

Gonostylus.

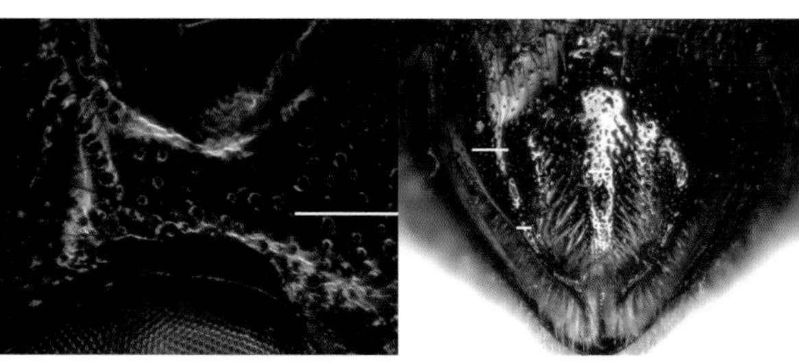

Inner eye margin.

Keel.

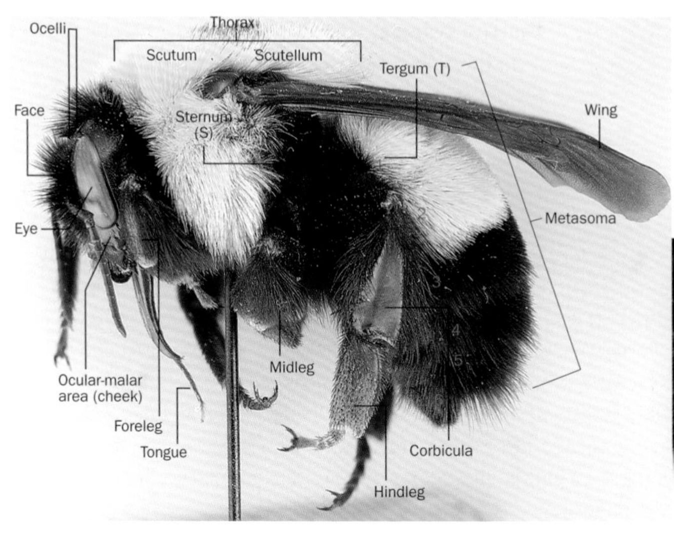

Body morphology.

BELOW: Left mandible.

Labrum The hinged upper lip, attached to the apex of the clypeus (see *clypeus*).
Lateral At the side.
Mandible The paired chewing outer mouthparts attached at the bottom end of the cheek (malar area).
Mandibular hinge The points of attachment of the mandibles to the head (see *mandible*).
Medial Toward the middle.
Mesosoma. The middle body part in Hymenoptera composed of the three thoracic segments and the first true abdominal segment (propodeum). Sometimes referred to loosely as the "thorax", even though it includes part of the abdomen.
Mesothorax, mesothoracic The middle thoracic segment, bearing the forewings and midlegs.
Metasoma The third body part in Hymenoptera, the part after the waist, composed of all abdominal segments except for the first (propodeum), which is included in the mesosoma. Sometimes referred to loosely as the "abdomen", even though it excludes one abdominal segment.
Midleg The leg of the mesothoracic segment.
Ocellus (pl. ocelli) One of the three simple eyes found at the top middle of the head that function to detect light rather than for image formation.
Pedicel The small second segment of the antenna attached between the scape and flagellum (see *antennal segments*).
Penis valve The paired median sclerotized (hardened) structures of the male genital capsule, used to clasp the female during mating.

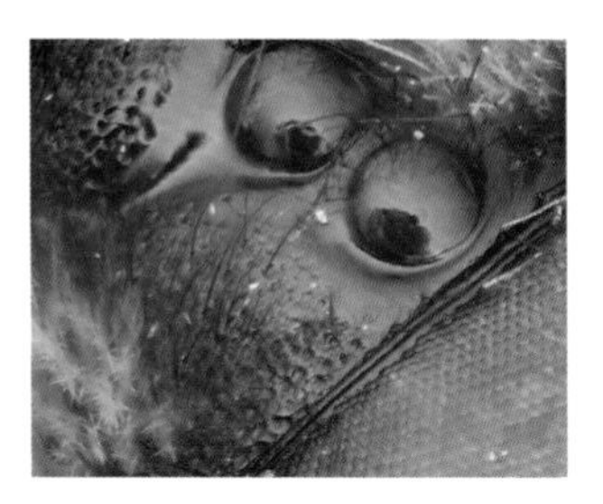

Mid- and left ocelli.

Penis valve.

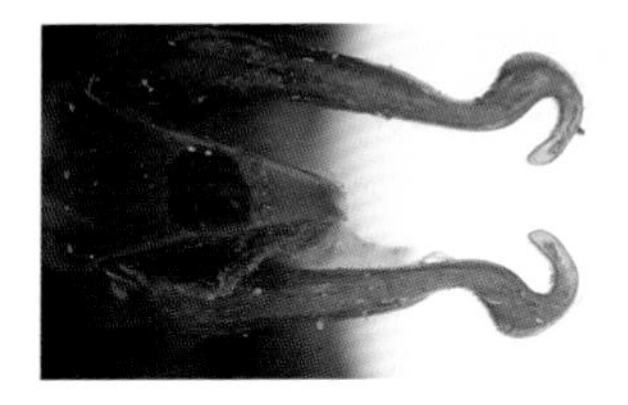

Penis-valve head.

Penis-valve head The part of the penis valve furthest from the body that is defined by an outer flange or ridge, the head often turned inward as a sickle shape or sometimes turned outward.

Posterior Toward the rear.

Process (e.g., proximal posterior process). A structure protruding from a surface, often toothlike or spinelike.

Propodeum The true first abdominal segment, fused to the thorax, together forming the mesosoma in aculeate (stinging) Hymenoptera, and separated from the rest of the abdomen (in the metasoma) by a constriction or waist.

Proximal Near to or adjacent to.

Scape The elongate basal segment of the antenna (see *antennal segments*).

Sclerotized Hardened, as in the rigid exoskeleton of an insect.

Scutellum The second, rearmost and smallest dorsal plate of the mesothorax, between the scutum and metanotum.

Scutum The first, anterior-most and largest dorsal plate of the mesothorax.

Spatha A median, dorsal, often nearly oval plate covering the penis.

Sternum (S#) A plate on the ventral or underside of the metasoma (loosely "abdomen"). The visible plates are numbered from the constriction between the mesosoma and metasoma starting with S1–6 visible externally in both the female and the male.

Submarginal groove (of the gonostylus) A longitudinal depression just below the inner margin of the gonostylus and running parallel to it (see *gonostylus*).

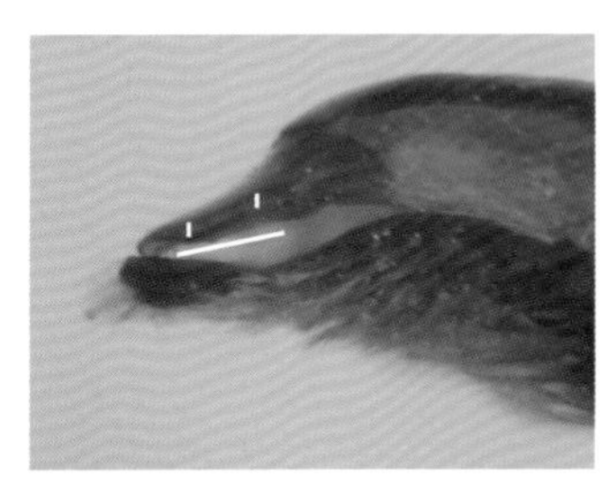

Submarginal groove.

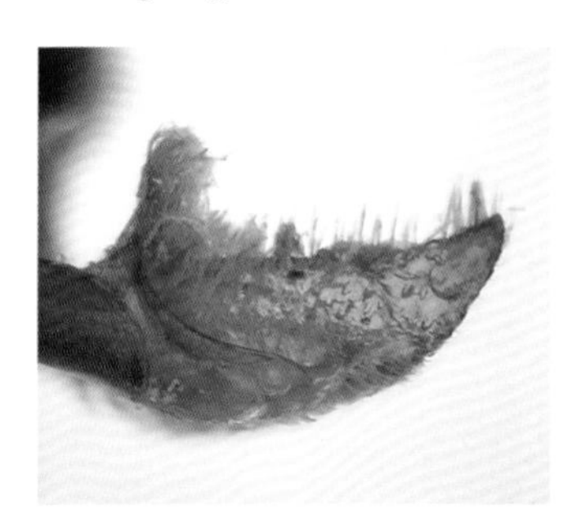

Volsella.

Tarsus (pl. tarsi) The furthest from the body of the five components of the insect leg (coxa, trochanter, femur, tibia, tarsus), forming a foot, and in bees itself composed of five "segments" or tarsomeres (see *basitarsus*).

Tergum (T#) A plate on the dorsal upper side of the metasoma. The visible plates are numbered from the constriction between the mesosoma and metasoma starting with T1–6 in the female or T1–7 in the male.

Thoracic dorsum The upperside of the thorax, mesosoma.

Tibia The part of a leg between the femur and tarsus, modified as a corbicula on the hindleg of nonparasitic female bumble bees (see *corbicula*).

Tooth An angled, pointed structure protruding from a surface.

Ventral Underside, below or at the bottom.

Volsella (pl. volsellae) Paired lateral structures of the male genitalia attached broadly near the inner base of the gonocoxa, often densely hairy, and protruding beyond the gonostylus in bumble bees, used to clasp the female during mating (see *gonostylus*).

ADDITIONAL RESOURCES

Befriending Bumble Bees: A Practical Guide to Raising Local Bumble Bees (2007) by Elaine Evans, Ian Burns, and Marla Spivak, University of Minnesota Extension. 65 pp.
BioQuip Products www.bioquip.com
Bugguide (*Bombus*) www.bugguide.net/node
Bumblebees: Behaviour, Ecology and Conservation (2010) by Dave Goulson, Oxford Biology. 336 pp.
Bumblebees and Their Ways (1934) by Otto Emil Plath, MacMillan. 201 pp.
Bumblebee Economics (2004) by Bernd Heinrich, Harvard University Press. 245pp. [Reissue of original 1979 edition with a new preface]
Bumble bees and cuckoo bumble bees of California (Thorp *et al.* 1983) (available online through Essig Museum of Entomology, UC Berkeley): http://essig.berkeley.edu/CIS/cis23.pdf
Bumble bees of the Eastern United States www.fs.fed.us/wildflowers/pollinators/documents/BumbleBeeGuideEast2011.pdf
Bumble bees of the Western United States www.fs.fed.us/wildflowers/pollinators/documents/BumbleBeeGuideWestern2012.pdf
Bumblebees of the World www.nhm.ac.uk/bombus/
Bumble Bee Watch www.bumblebeewatch.org
Discover Life www.discoverlife.org/mp/20q?guide=Bumblebees
Keeping the Bees (2010) by Laurence Packer, HarperCollins. 272 pp.
NatureServe www.natureserve.org
North American Pollinator Protection Campaign nappc.org
Pollinator Partnership www.pollinator.org/
Pollinator Partnership Canada www.pollinatorpartnerhsip.ca
Rose Entomology Supplies www.roseentomology.com
The Bumblebee Conservation Trust www.bumblebeeconservation.org.uk/
The Great Sunflower Project www.greatsunflower.org/
The IUCN Red List of Threatened Species www.iucnredlist.org
The Natural History of Bumblebees: A Sourcebook for Investigations (2001) by Carol A. Kearns and James D. Thomson, University Press of Colorado. 130 pp.
The Xerces Society for Invertebrate Conservation www.xerces.org/bumblebees/
Wildlife Preservation Canada www.wildlifepreservation.ca/

ACKNOWLEDGMENTS

We would like to thank our friends, family, and colleagues who supported us throughout the creation of the book.We are particularly grateful to the following individuals who provided substantial assistance to various aspects of development of this book: Lucy Bailey, Michael Kuhlmann, Vlad Blagoderov, Eugene Morton, Sam Droege, Steve Buchmann, Sarina Jepsen, Laurence Packer, Cory Sheffield, Jonathan Chipman and Russell Galen. Thank you to our reviewers Syd Cannings, Joe Engler, Terry Griswold, David Walter and Stephen Buchmann, as well as one anonymous individual. We thank our institutions for providing space and materials required to gather the information presented in this guide: Natural History Museum, London; University of California, Davis; York University, Toronto; and Dartmouth College, Hanover. Many thanks to Jonathan Chipman for help with map-data conversions and Ann Sanderson for help with activity-plot graphics file conversions.

We are very grateful to the following individuals and collections which provided more than 200,000 specimens from which much of the information presented in this guide was gathered. We are also grateful to the hundreds of collectors, lab technicians, and collection managers crucial to the preservation and curation of these historical records: Academy of Natural Sciences, Philadelphia; Algonquin Provincial Park Insect Collection; André Francoeur, University of Quebec; Andrew Hebda, Nova Scotia Museum; Ball State University, Indiana; Barr Entomology Museum, University of Idaho, Moscow; Bernd Heinrich, University of Vermont; Bernice Pauahi Bishop Museum; Brigham Young University; British Natural History Museum; California Academy of Sciences; California State Collection of Arthropods; Canadian Museum of Nature; Canadian National Collection; Caroline Gottschalk, University of Rhode Island; Caroline Scully; Cathy Stragar, University of Delaware; Christophe Buidin/Yann Rochepault research collection; College of Idaho; College of the North Atlantic, Carbonear; Comisión Nacional para el Conocimiento y Uso de la Biodiversidad, Mexico; Connecticut Agricultural Extension Station; Cory Sheffield, Royal Saskatchewan Museum; Dan Fiscus, Frostburg University; Death Valley National Park; DePauw University; Derek Sikes, University of Alaska, Fairbanks; Donald Miller, Lyndon State College; Doug Gill, University of Maryland; Doug Golick; Doug Yanega, University of California, Riverside (NSF Grant DBI-0956340); Elaine Evans, University of Minnesota; Fawziah Gadallah, Environment Canada; Field Museum of Natural History (Chicago); Fred Morrison; Fundacion Universidad de las Américas Puebla; Gerry Stage; Heather Axen and Kurt Pickett, University of Vermont Zadock Thompson Zoological Collection; Heather Hines, North Carolina State University; Helen Young, Middlebury College; Ignasi Bartomeus and Rachel Winfree, Rutgers University; Illinois Natural History Survey; Indiana Dunes Entomology Collection; Indiana State University Terre Haute; James Strange, Terry Griswold, and Jonathan Koch, University of Utah, Logan; James Thomson, University of Toronto; Jamie Cromartie, Stockton College; J. B. Wallis Museum of Entomology, University of Manitoba; Jeff Freeman, Castleton State College; Jeff Ogden, Nova Scotia Department of Natural Resources; Jennifer Thomas, University of Kansas; Joan Milam, University of Massachusetts; Joel Gardner, University of Minnesota; John Ascher (NSF Grant DBI-0956388), American Museum of Natural History; Karen Wetherill, University of New Mexico; Kent McFarland, Vermont Center for Ecostudies; Kevin Matteson, Fordham University; Kyle Martins, McGill University; Laval University; Lawrence Gall, Yale University; Lethbridge Research Centre, Agriculture and Agri-Food Canada; Liz Day; Los Angeles County Museum; McGill University, Lyman Entomological Collection;

Michael McKelvey, Beespotter; Michael Veit, Lawrence Academy; Michael Warriner, Texas Parks and Wildlife Department; Michel Savard; Middlebury College; Mike Arduser, Missouri Department of Conservation; Mississippi Entomological Museum; Missouri Department of Natural Resources; Molly Notestine; Natural History Museum, Bern, Switzerland; Neal Williams, University of California, Davis; New York State Museum; North Carolina State University; Nova Scotia Department of Natural Resources; Nova Scotia Museum; Ohio State University; Oregon State University Corvallis Arthropod Collection; Packer Collection, York University; Patuxent Wildlife Research Center; Peter Hallett; Prince Edward Island Museum; Purdue University; Rebecca Irwin, Dartmouth College; Rich Hatfield and Sarina Jepsen, Xerces Society; Rob Jean, Saint Mary of the Woods College; Robert Parmenter, Valles Caldera National Preserve; Rocky Mountain Biological Laboratory; Ross Bell; Royal Ontario Museum; Sam Droege, USGS Patuxent Wildlife Research Center; San Diego Natural History Museum; San Diego State University; Sara Zahendra, Vermont Center for Ecostudies; Sherry Surrette, Mississippi State University; Smithsonian Institution; South Dakota State University; Staten Island Institute of Arts and Sciences; Steve Javorek, Agriculture and Agri-Food Canada; Stockton College; Sue Sheehan, Fermilab; Sydney Cameron, University of Illinois, Urbana-Champaign; Terence Schiefer, Mississippi State University; Texas A&M University; Texas Memorial Museum Austin; Universidad Nacional Autonóma de Mexico; University of Alaska Fairbanks; University of Arizona; University of British Columbia Vancouver; University of California Berkeley Essig Museum; University of California Davis Bohart Museum; University of Colorado Boulder; University of Connecticut Storrs; University of Guelph; University of Kansas Snow Entomology Museum; University of Massachusetts Amherst; University of Minnesota; University of Montana Bozeman; University of Nebraska; University of Nevada, Reno; University of New Hamphsire, Durham; University of New Mexico (Sevilleta Long Term Ecological Research); University of Notre Dame Museum of Biodiversity; University of Regina; University of Washington, Seattle; University of Wisconsin Madison; University of Wyoming; Vermont Department of Forests, Parks and Recreation; Victoria MacPhail; Virginia Scott, University of Colorado, Boulder; Washington State University Pullman. Special thanks to John Ascher and Doug Yanega for the databasing project that made the locality and phenology data possible.

PHOTO CREDITS

We thank and acknowledge the contributions of numerous amateur and professional photographers. Without them and their beautiful photographs, this book would have not have been possible.

AB—Anne Brigham
AF—Alistair Fraser
AFO—Ashley Fortney
AR—Andrew Richards
AS—Ann Sanderson
BH—Bernd Heinrich
BK—Brian Klinkenberg
CB—Curtis Bjork
CD—Carol Davis
CS—Christy Stewart
CSH—Cory Sheffield
CW—Charles Wead
DG—Dave Goulson
DI—David Inouye
DK—Donald Kramer
DM—Dan Mullen
DR—Don Rolfs
DRU—Deborah Rudis
DT—Doug Tate
GH—Göran Holmström
GZ—Gary Zamzow
HB—Heinz Baum
HP—Heather Proctor
HW—Hartmut Wisch
JCJ—J. C. Jones
JD—Jason Drombroskie
JE—Joe Engler
JF—Jessica Forrest
JG—Joel Gardner
KKG—Kathy Keatley Garvey
LL—Liana Lewis
LR—Leif Richardson
LRE—Larry Rea
LW—Larry Watkins
MH—May Haga
MM—Margarita Miklasevskaja
NO—Nick Owens
NS—Nick Saunders
PH—Pat Hinds
PW—Paul Williams
RB—Robert Bercha
RC—Rollin Colville
RH—Ron Hemberger
RHA—Rich Hatfield
RHE—Robert Herr
RO—Robin Owen
ROD—Ryan O'Donnell
RP—Rehanon Pampell
RT—Rosemary Taylor
RTH—Robbin Thorp
RW—Robin Williams
SC—Syd Cannings
SCA—Suzanne Carriere
SCAR—Susan Carpenter
SCO—Sheila Colla
SD—Sheila Dumesh
SF—Steven Falk
SR—Samina Rutrecht
TB—Tom Bentley
TJ—Tim Johnston
TL—Tim Lethbridge
TS—Tim Schowalter
VK—Vicky Kjoss
VL—Veronica Ladico
VM—Victoria MacPhail
WE—Werner Eigelsreiter

INDEX